GO FIGURE

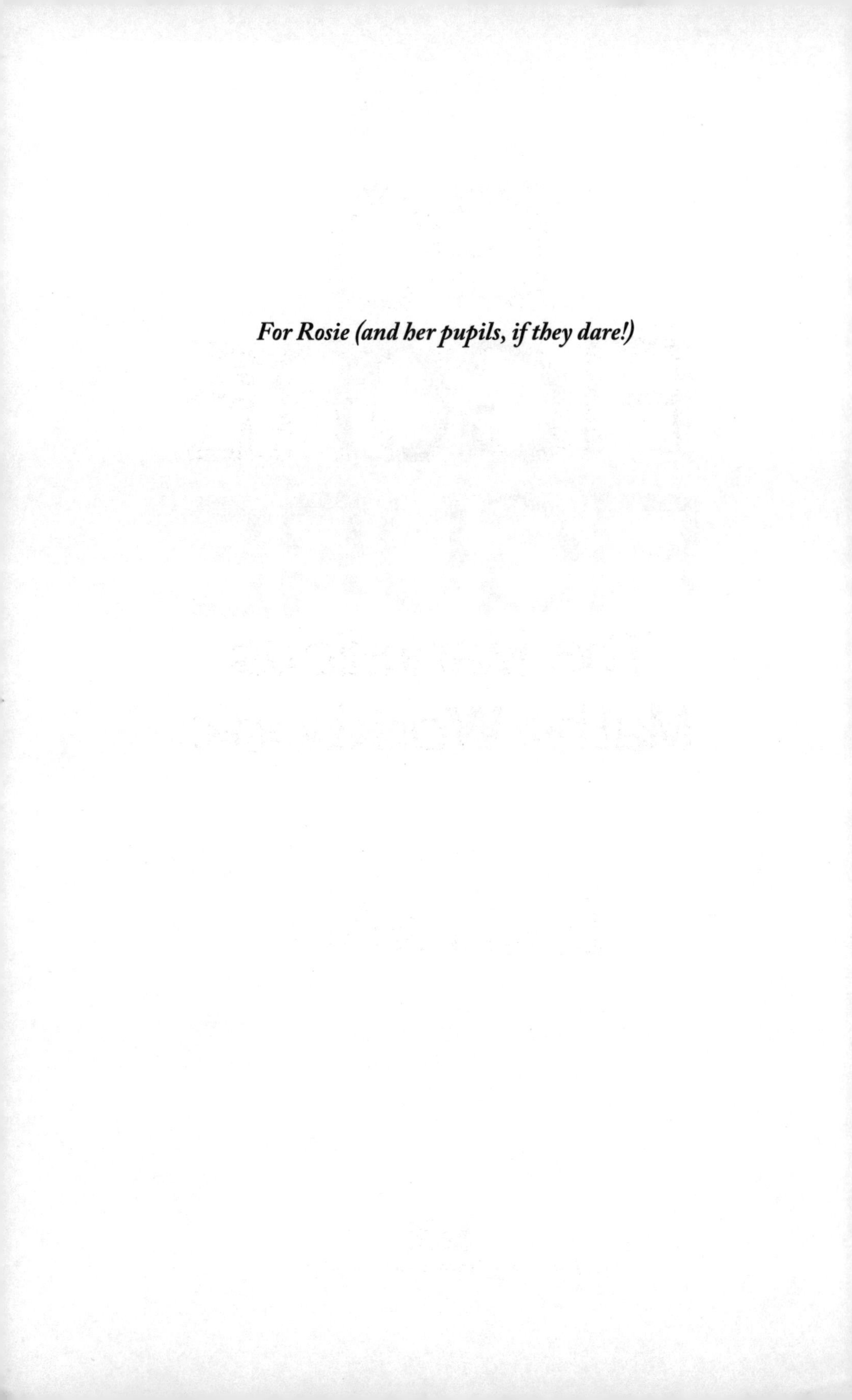

For Rosie (and her pupils, if they dare!)

GO FIGURE

The Marvellous Maths Workbook

Daniel Smith

Michael O'Mara Books Limited

First published in Great Britain in 2012 by
Michael O'Mara Books Limited
9 Lion Yard
Tremadoc Road
London SW4 7NQ

A CIP catalogue record for this book is available from the British Library.

Papers used by Michael O'Mara Books Limited are natural, recyclable products made from wood grown in sustainable forests. The manufacturing processes conform to the environmental regulations of the country of origin.

ISBN: 978-1-84317-864-4

1 3 5 7 10 8 6 4 2

Cover design by Anna Morrison
Designed and typeset by www.glensaville.com
Technical drawings by Greg Stevenson
Illustrations by Andrew Pinder

Printed and bound in Great Britain by Clays Ltd, St Ives plc

www.mombooks.com

CONTENTS

INTRODUCTION

For those who 'get' maths, there is little else more perfect that the human mind can imagine. Indeed, it is often said that our understanding of the world around us (and the universe as a whole, for that matter) would be nought without maths and that it is the purest of all the sciences.

But what of the far greater number of people for whom maths comes bound up in fear and trepidation? How many of us sat in musty schoolrooms never able to conquer the 9 times table or understand what the point of a square on a hypotenuse is anyway? How many handed in their last piece of enforced maths homework vowing never to engage with the subject in any meaningful way again? And how many of us still have the odd nightmare involving turning up for a maths exam without a clue as to how to do any of it?

The truth is that these conflicting points of view can and do coexist. It is a fact that our greatest minds would have little comprehension of the big questions of time and space were it not for maths. Equally, it would be all but impossible to buy something in a shop or tell the time or observe a speed limit or know when our next birthday is without a sound grasp of some rudimentary maths principles. Every day, we use vast numbers of mathematical skills and techniques, often without even realizing it.

But reassuringly, even the greatest minds are sometimes flummoxed by the subject. Take the words of the great British mathematician and philosopher of the last century, Bertrand Russell. In 1910 he wrote: 'Mathematics, rightly viewed, possesses not only truth, but supreme beauty – a beauty cold and austere, like that of sculpture.'

Eight years later, he added these thoughts (after struggling through a quadratic equation, I like to imagine): 'Mathematics may be defined as the subject in which we never know what we are talking about, nor whether what we are saying is true.'

And so we come to this book, which I hope will be an assistance and (say it quietly) perhaps even an entertainment to those who know the value of maths and want to get to grips with some of its intricacies.

Contained herein are dozens of quizzes designed to help you take on different key topic areas in easily digestible chunks. You will get to try your hand at everything from the simplest addition sum to long division and probability theory, from simultaneous equations to scattergraphs and trigonometry.

Some of the quizzes you should be able to get to grips with relatively easily. Others are likely to occupy the little grey cells considerably more, but there is nothing in here that need be beyond your reach. Each section comes complete with some introductory notes to guide you through. Occasionally, you will need to employ a calculator but most quizzes can be answered with only a pencil and brain power (even if some of them don't look like that at first). Don't consider the calculator as a symbol of defeat if you do need to pick it up but, equally, don't be tempted to use one when you don't really need to.

Ultimately, this book aims to make maths approachable. Whether you're currently studying the subject or dipping your toes back in after many years away, treat the quizzes as an opportunity to brush up your maths know-how while having a bit of fun. Embrace the challenge, give yourself a pat on the back when you conquer a quiz and don't get disheartened if a particular question defeats you. After all, not even Bertrand Russell always knew what he was talking about…

IT ALL ADDS UP: ARITHMETIC

Arithmetic is the term we use to describe all those basic operations we do every day with two or more numbers; all those simple processes of addition, subtraction, division and multiplication that, for instance, help us plan our days (what time will it be in three hours?), pay for our groceries (can I afford a pint of milk, a loaf of bread and a newspaper with the £2 in my pocket?) and even keeping our families happy at Christmas (if I want to give each of my nephews three handkerchiefs, how many must I buy?).

Simple Symbols

Mathematics has its own language, consisting not only of numbers but of symbols as well. The first quiz contains several of the most commonly used symbols along with explanations of their functions.

Quiz 1

Can you match up each symbol to its correct definition?

1) +	a) infinity
2) −	b) is less than or equal to
3) ×	c) does not equal
4) ÷	d) pi (used in numerous calculations related to circles)
5) =	e) square root
6) ≠	f) is greater than
7) ≈	g) minus
8) <	h) divided by
9) >	i) is less than
10) ≤	j) equals
11) ≥	k) multiplied by
12) √	l) is approximately equal to
13) ∞	m) plus
14) π	n) is greater than or equal to

Answers on page 163

All in the Mind

Mental arithmetic is not the name for sums that drive you crazy but refers to those bits of maths that you do in your head, sometimes without even thinking about it. Most mental arithmetic involves smallish numbers that are easier for our grey cells to handle, rather than complex calculations involving figures the length of telephone numbers.

Quiz 2

Let's start with a few simple mental arithmetic testers to get those mathematical cogs whirring. You shouldn't spend more than a few seconds on each question.

1) 9 + 7 =
2) 22 – 14 =
3) 6 x 4 =
4) 17 + 9 =
5) 28 ÷ 4 =
6) 8 x 5 =
7) 38 – 19 =
8) 27 + 17 =
9) 8 x 7 =
10) 45 ÷ 9 =

Answers on page 164

Sign of the Times

The bane of many a school child's life, learning your times tables until you know them back to front is a great skill to have and will make all sorts of day-to-day calculations that bit easier. There are even a few simple rules that can help you:

- *For numbers in the 2 times table, you are simply doubling. (e.g. 2 x 3 = 6)*
- *For numbers in the 4 times table, you are doubling twice (e.g. 4 x 3 = double double 3 = double 6 = 12)*
- *For numbers in the 8 times table, you are doubling three times (e.g. 8 x 3 = double double double 3 = double double 6 = double 12 = 24)*

- *For numbers in the 5 times table, multiply by ten and then halve the result (e.g. for 5 x 3, calculate 10 x 3 [which is 30] and then halve it; 30 ÷ 2 = 15).*

- *For numbers in the 10 times table, add a 0 (e.g. 10 x 7 = 70)*

- *In the 11 times table, simply repeat the number you're multiplying by for figures up to 9 (e.g. 11 x 3 = 33). We know how to calculate 10 x 11 (see above), but 11 x 11 and 12 x 11 you'll just need to learn by heart.*

- *For the 9 times table, you can use your hands for figures up to ten. Here's how... Put both hands out in front of you. From the left, count to the finger that represents the number of nines you want (so for 3 x 9, count to the middle finger of your left hand). Fold that finger down. This leaves 2 fingers before it and 7 after it, so 3 x 9 = 27. Magic! We know how to work out 9 x 11 from the 11 times table rule above, which means you only need to memorize 12 x 9.*

Quiz 3

Armed with these tricks of the trade, try tackling the multiplication table below. See how quickly you can do it (and no cheating!).

1	2	3	4	5	6	7	8	9	10	11	12
2	4										
3											
4			16								
5											
6											
7											
8											
9											
10											
11											
12											

Answers on page 164

Pluses and Minuses

A negative number is one that is worth less than 0. You can spot them because they have a minus symbol before them. They are a part of pure maths, as opposed to applied maths, which is to say that they are theoretical rather than actual. We cannot, for instance, hold –2 apples in our hands.

You might wonder when you need to talk about numbers less than 0 but they are surprisingly useful. For instance, if you are trying to get on top of your finances, you may well find yourself needing to do some calculations incorporating all those negative numbers on your bank statement!

Because negative numbers carry a minus sign with them all the time, working out how they fit into addition and subtraction sums can be pretty taxing:

$3 + 4 = 7$
$-3 + 4 = 1$
$-3 - 4 = -7$
$3 - 4 = -1$
$3 - -4 = 7$

Got all that? I thought not! It might help to visualize the numbers on a number line like the one below:

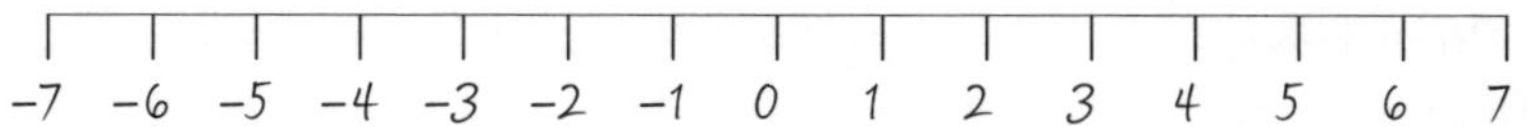

Imagine the first number in a sum is your starting point. Go and stand by that number. The second number is your guide to where to go next.

If your sum asks you to add something, turn to face the higher numbers. If it's a subtraction, face the lower numbers. If the number you're adding or subtracting is positive, carry on in the direction that you're facing. If it's a negative number, walk backwards.

Take that last sum as an example. Position yourself at 3 on the number line. It's a subtraction, so face the lower numbers. The second figure is negative so walk backwards 4 paces. Where do you end up? At 7. Job done (and proof that two negatives make a positive, just like your old schoolmaster always told you).

Quiz 4

Time for you to try a few of your own:

1) $6 + -4 =$
2) $-3 - 8 =$
3) $-5 + 9 =$
4) $8 - 13 =$
5) $-3 - 4 - 5 =$
6) $-17 + 13 =$
7) $4 + -7 =$
8) $-8 + 17 =$
9) $-5 - 21 =$
10) $-3 - -8 =$

Answers on page 165

Turning a Negative into a Positive

Multiplying a positive number by a negative number always results in a negative figure: -2×4, for instance, comes to -8. However, multiply 2 negative numbers together and you get a positive: $-2 \times -4 = 8$.

The strangeness doesn't stop there. Normally when we divide, we end up with a smaller number than we started with. For instance, if we divide 10 by 2 we get 5. But if we divide –10 by 2 we get –5 (and –5, of course, is a bigger number than –10).

In short, if you multiply or divide numbers that are either both negative or both positive, your answer will be positive. If one number in your sum is positive and one negative, the answer will be negative.

Quiz 5

1) –3 x 4 =
2) –6 x –7 =
3) –38 ÷ 2 =
4) 3 x –28 =
5) –27 ÷ –3 =
6) –120 x –7 =
7) –144 ÷ 24 =
8) 931 ÷ –7 =
9) –42 x –15 =
10) (156 ÷ –13) + 6 =

Answers on page 165

Everything in its Place

Every digit within a number represents a particular value. What that value is depends on which 'column' within a number the digit is in (i.e. its 'place value').

If we are dealing with whole numbers, the columns increase in value from right to left. So the figure farthest to the right represents the number of units we are dealing with, the column to its left tells us how many 10s, to its left come the 100s, to its left the 1,000s, to its left the 10,000s, to its left the 100,000s, to its left 1,000,000s and so on.

Take, for instance, the number 1,234. What can we learn from this? Well, reading from right to left, we see that the number comprises 4 units, 3 lots of 10 (e.g. 30), 2 lots of 100 and 1 lot of 1,000.

Now, if we have a number with a decimal point, we read the figures after the decimal point from left to right. So the first column tells us how many tenths we are dealing with, the column to its right how many hundredths, to its right how many thousandths and so on.

Quiz 6

1) Which digit represents 'tens' in the following number: 1,364?
2) Which digit represents 'tens of thousands' in the following number: 27,345,962?
3) Which digit represents 'tenths' in the following number: 9.783?
4) Which digit represents 'millionths' in the following number: 1.23456789?
5) What is 27 x 100?
6) What is 36,275,000 ÷ 100?
7) What is 28.75 x 1000?
8) What is 1.3825 ÷ 100?

Answers on page 166

Paperwork

When calculations are too complicated to do in your head (or on your fingers), we enter the realm of paper arithmetic. Normally, this means using column arithmetic, which is where having a good understanding of place value comes into its own. This is because column arithmetic needs us to line up all the units, tens, hundreds, thousands, etc. in the numbers we are working with.

And Another Thing

To add two or more numbers using paper arithmetic, we put them in a tower with all the columns correctly aligned. It's then a case of doing the simple addition of the numbers in each column from right to left and putting the result at the bottom. For instance:

```
 1342 +
  257
 ----
```

So first we add the 7 + 2 in the last column (=9), then 5 + 4 in the next one along (=9), then 3 + 2 (=5) and, finally, 1 + 0 (=1).

```
 1342 +
  257
 ----
 1599
```

This works perfectly until the numbers in any column add up to more than 9. If this occurs, we have to start 'carrying'. All this means is that if you end up with a 'spare' digit, you move it along to the column to the left. Say you are adding 169 + 87:

```
 169 +
  87
 ---
```

Well, 9 + 7 = 16, so we put 6 at the bottom of the column on the far right and carry the 1 (in reality, a value of ten) to the 'tens' column to its left. You can put the spare figure at the top of the column to remind you it needs to be added. Then we add the 8, the 6 and the spare 1. This gives us 15, so 5 goes at the bottom of the middle column and the spare 1 goes into the last column of the left, which when added to the 1 already there gives us 2. So our answer is 256:

$$\begin{array}{r} {}^{1}\,{}^{1} \\ 169 \\ +\;\;87 \\ \hline 256 \end{array}$$

Quiz 7

Time to try a few of your own:

1) Dave, who liked to keep count, received 43 birthday cards and was very hurt when his wife, Sheila, received 88. How many did they get in total?
2) Kevin worked 36 hours in the first week of a new job, 41 hours the following week, 48 the week after that and 35 in week 4. How many hours did he work over the month?
3) Two local schools took their pupils to the theatre. The first school had 662 students while there were 955 in the second school. How many pupils went to the show in total?
4) A census was taken of three villages. Lower Snodgrass had a population of 552, Upper Whimsy boasted 811 residents and Keinton Widdlethorpe had 1,214 locals. What was their combined population?
5) Clive has three bank accounts. In Bank A, he has £1,352. Bank B has £2,889 of his money. He has deposited £4,651 at Bank C. How much does he have in the three accounts combined?

Answers on pages 166–7

Take It Away

When subtracting on paper, the number you are taking away should be underneath the number you are taking away from. Then, in the same way as addition, start at the farthest column to the right and work back, in each case subtracting the lower figure from the one above it. But what to do if the number you're subtracting is bigger than the one you're subtracting from? This is when we need to 'borrow', which is essentially 'carrying' in reverse.

Take the example 349 – 87:

```
 3 4 9 _
   8 7
 -----
 ? ? ?
```

Starting from the far right column, 9 – 7 = 2. So far, so good. But next, in the tens column, we have 4 – 8. That is -4 but we really don't want to start chucking negative numbers into our result. So we 'borrow' a 1 from the hundreds column. Now we have 14 – 8, which is an altogether more convenient 6. But having borrowed 1 from the 3 in the hundreds column we must remember to change the 3 to a 2. That gives us 2 – 0 = 2. So our result is 262:

```
 3̸2 ¹4 9 _
     8 7
 -------
  2  6 2
```

Oh, and remember, if you need to you can borrow from a '0', the 0 becomes a 9 but you'll also have to borrow from the next whole number (and any other 0s in between). So:

```
 3 0 0 5 _
   1 8 9
 -------
```

Becomes:

$$
\begin{array}{r}
\not{3}2\ {}^{1}\not{0}9\ {}^{1}\not{0}9\ {}^{1}5 \\
1\ \ 8\ \ 9 \\
\hline
2\ \ 8\ \ 1\ \ 6
\end{array}
\;+
$$

Quiz 8

1) If Sheila received 97 Christmas cards, and Dave got 68, how many fewer did he receive?
2) Kevin worked 137 hours in one month but only 99 hours the next. How many fewer hours did he work in month 2 than in month 1?
3) The theatre had 2,027 seats, of which two local school parties took up 1,617 seats. How many seats were left?
4) Three villages in a borough had a combined population of 2,577 residents. However, owing to border changes, Upper Whimsy with 811 residents became part of the neighbouring borough. How many people lived in the other two villages combined?
5) Clive had 3 bank accounts with combined deposits of £8,892. He then took out £3,973. How much did he have left banked?

Answers on pages 167–8

Go Forth and Multiply

The easiest way to get to grips with multiplication is to think of it as doing an addition but lots of times. So 3 x 9 is the same as adding together 9 lots of 3. However, that takes quite a lot of time so multiplication is your quick route to the answer.

The trick is to make all the sums you need to do as simple as possible. If someone asks you what 9 x 14 equals, your brow might furrow. But if you treat it as two separate calculations which you then add together (e.g. 9 x 10 = 90 and 9 x 4 = 36; 90 + 36 = 126), it doesn't seem quite as scary.

Take another even more complicated sum, such as 27 x 31. Divide it into simpler 'chunks':

20 x 30 = 600
20 x 1 = 20
7 x 30 = 210
7 x 1 = 7
Total = 837

Some people find it easier to use a grid:

	30	1
20	20 x 30 = 600	20 x 1 = 20
7	7 x 30 = 210	7 x 1 = 7

Quiz 9

Have a go at solving the following questions using the grid method:

1) 17 x 8 =
2) 16 x 13 =
3) 34 x 15 =
4) 42 x 26 =
5) 177 x 52 =

Answers on pages 168–9

Times and Times Again

Then there is long multiplication! This usually involves lots of 'carrying', as we had to do with some of our addition calculations. Let's calculate 29 x 128 as an example. First, set out the sum like this:

```
128 x
 29
____
```

As ever, start in the far right column. We know from our times tables that 9 x 8 = 72. So 2 goes at the bottom of the column and 7 gets carried over to the tens column. Now multiply 9 by the 2 in the tens column and then **add** the 7 you've carried over to the result (which gives 25, so 5 goes at the bottom of the column and the 2 is carried over). Next, multiply 9 by the 1 in the hundreds column and add the 2 you've carried over, like this:

```
 27
 128 x
  29
 ----
1152
```

OK, that's half the calculation done. Now to use the 2 in 29. This time put your results underneath the 1152. But remember, the 2 in 29 is in the tens column, so it's really a 20. For this reason, you won't be putting any numbers in the units column. Fill it up with a zero to avoid the temptation. Now on with the maths. Anything you need to carry over this time should go above the 27.

2 x 8 = 16. So 6 goes in the tens column and 1 gets carried. Next, do 2 x 2 and add the carried 1. The result is 5. Finally, you're left with 2 x 1 = 2.

```
  1
  ~~27~~
  128 x
   29
 ----
 1152
 2560
```

Then it's simply a case of adding your two figures together, which comes to 3,712.

(Remember, if the figure you're multiplying by has three figures, you will need to put 0s in the units and tens columns when multiplying by the figure in the hundreds columns, and add an additional zero for every additional row of results).

Quiz 10

1) Fred saved 16 pounds per week for a whole year. How much has he saved by 31 December?
2) Des had 17 full stamp albums, each with enough room for 121 stamps. How many stamps does Des have in total?
3) Mark has 25 employees who each earn £187 per week. How much money does Mark pay out each week in wages?
4) 487 people attended the concert, each paying £38 for a ticket. How much money did the ticket office take?
5) With 24 hours in a day, how many hours are there in a non-leap year?
6) If there are 60 seconds in a minute, 60 minutes in an hour and 24 hours in a day, how many seconds in a day are there?

Answers on pages 169–71

Divide and Rule

Of all the basic mathematical functions, division tends to be the one that most frightens people but it needn't be terrifying. It sometimes helps to rephrase a 'divide' question into a multiplication one. For instance, the sum 42 ÷ 6 is asking you what 42 is when split into 6 equal parts. Having become a master of the times tables, you know that 6 goes into 42 a total of 7 times, so 7 is your answer.

You can bring your knowledge of the times tables to bear in more complex calculations where the number you're dividing by is 12 or less. Take the sum 204 ÷ 12. Write it like this:

$$12\overline{)2\ 0\ 4}$$

Work from the left hand column towards the right. 12 doesn't go into 2, so put 0 above the 2 and carry the 2 across to the next column:

$$\begin{array}{r|l} & 0 \\ \cline{2-2} 12 & 2\ {}^{2}0\ 4 \end{array}$$

12 goes into 20 once with 8 left over. So put 1 in your answer and carry the 8 into the next column. Then the final calculation is 12 into 84:

$$\begin{array}{r|l} & 0\ \ 1\ \ 7 \\ \cline{2-2} 12 & 2\ {}^{2}0\ {}^{8}4 \end{array}$$

So 204 ÷ 12 = 17.

Of course, sometimes one figure doesn't divide into another one exactly. In such cases, you end up with a remainder in your final column. Fear not! There is a method to deal with such scenarios. Simply add a decimal point after the final figure in the number you're dividing into and add a 0 after it. Carry on your calculations as before, adding more 0s after the decimal point until you have found your answer. Here's an example using 116 ÷ 8:

$$\begin{array}{r|l} & 0\ 1\ \ 4\cdot 5 \\ \cline{2-2} 8 & 1\ {}^{1}1\ {}^{3}6\cdot{}^{4}0 \end{array}$$

So 116 ÷ 8 = 14.5.

One other quick note. Every now and again, a division sum gets caught in a recurring loop. Try dividing 1 by 3:

$$\begin{array}{r|l} & 0\ .\ 3\ 3\ 3\ 3\ 3\ 3\ 3\ 3 \\ \cline{2-2} 3 & 1\ .\ {}^{1}0\ {}^{1}0\ {}^{1}0\ {}^{1}0\ {}^{1}0\ {}^{1}0\ {}^{1}0\ {}^{1}0 \end{array}$$

No matter how decimal places you go to, the sum will go on like this. It is called a recurring decimal and you can signify it in shorthand simply by putting a dot over the recurring figure. If it is a sequence of numbers that recurs, a dot goes over the first and last digit.

Quiz 11

Now it's your turn.

1) John has £84 to last him for 7 days. How much money can he spend each day on average?
2) The record producer had recorded 192 tracks. If he can fit 12 tracks on an album, how many albums could he fill up?
3) In the raffle, £275 of gift vouchers was shared by 10 winners. How much did each winner get?
4) Sid has 198 grams of tuna in his fridge, in 9 equally sized cans. How much tuna is there in each can?
5) There are 174 children in the school, divided into six classes of equal size. How many pupils are there in each class?
6) Eight keen curtain-makers are allowed to share 300 metres of fabric. How many metres does each go home with?
7) On a mountaineering exhibition, Kenny Crampon has 8 Kendal mint cakes to last him 3 days. Assuming he eats equal amounts each day, how many bars will he consume every 24 hours?

The Long Way Round

For more complicated division sums, we turn to long division. Now, in the age of the calculator, there are plenty of maths teachers quite prepared to wave goodbye to long division altogether. But you're not going to get away with it that easily…

In case all that business with remainders and 'bringing down' is not as clear in your mind as it once was, let's go through a specimen example: 608 ÷ 19

Answers on pages 171–2

Step 1:
First off, write the problem as follows:

$$19\,\overline{)608}$$

Step 2:
Reading from the left, we know 19 doesn't go into 6 but it can go into 60.

$$\begin{array}{rl} & \;3 \\ 19 & \overline{)608} \end{array}$$

Step 3:
OK, so now we multiply the 3 by the 19, giving us 57, which we put under the 60.

$$\begin{array}{rl} & \;3 \\ 19 & \overline{)608} \\ & \;57 \end{array}$$

Step 4:
Now we subtract the 57 from 60, to give us 3.

$$\begin{array}{rl} & \;3 \\ 19 & \overline{)608} \\ & \;57 \\ \hline & \;\;3 \\ \hline \end{array}$$

Step 5:
Next we move down the spare 8 from 608 and put it next to the 3.

$$\begin{array}{rl} & \;3 \\ 19 & \overline{)608} \\ & \;57 \\ \hline & \;38 \end{array}$$

Step 6:
Now we see if 19 can go into 38… and it can. Twice. So we put a 2 into our result.

```
     32
19 | 608
     57
     --
     38
```

There is no remainder and nothing left to bring down, so we have our answer… 32!

Quiz 12

Still confused? See how you manage with these three progressively more testing teasers.

1) 156 ÷ 12 =
2) 837 ÷ 31 =
3) 10,701 ÷ 87 =

Answers on pages 172–3

In Your Prime

A prime number is any number which is greater than 1 but divisible only by 1 and itself. There are infinite numbers of prime numbers and they are of immense use in theoretical maths. However, here we will only be using a few of the 168 primes with a value under 1,000.

Quiz 13

Let's start off by listing all the factors of six random numbers. A factor is any number that divides exactly into your base number. Having done that, identify which of the six random numbers are also prime numbers.

1) 2
2) 4
3) 7
4) 9
5) 11
6) 21

Answers on page 173

Quiz 14

Now let's see if you can spot the prime in each of the following groups.

1) 15, 17, 21, 27
2) 33, 35, 37, 39
3) 83, 85, 87, 91
4) 405, 407, 411, 419
5) 711, 723, 733, 741

Quiz 15

Finally, look at the groups of numbers below (all of which are prime). Can you continue the sequence?

1) 5, 7, 11, __
2) 17, 19, 23, __
3) 73, 79, 83, __
4) 401, 409, 419, ___
5) 977, 983, 991, ___

Everything in Proportion

A ratio is a comparison of two numbers, often denoted by a colon separating them. For instance, if you have a bag containing 3 apples and 5 oranges, the ratio of apples to oranges could be represented as 3:5. If you always bought fruit in that ratio (for reasons best known to yourself), and on a given day we knew you had 9 apples in your bag, we would know you were also carrying 15 oranges.

Answers on pages 173–4

Meanwhile, a proportion is an equation showing that two ratios are equal. For instance, imagine you are building a little wall at one end of a garden that has the same proportions as a larger wall at the other end. The large wall is 3 metres wide and 1.5 metres tall, and the width of the smaller wall is 1 metre. We might represent the proportion like this: 3/1.5 = 1/?. A quick calculation reveals that to keep the proportion balanced,? must equal 0.5 metres.

Quiz 16

1) In a horse race, 6 horses were grey and 8 were brown. What is the ratio of grey horses to brown expressed in its simplest form?
2) The ratio of girls to boys in a school is 13:12. If there are 144 boys in the school, how many girls are there?
3) A gardener has an allotment that he has divided into 2 parts in the ratio 3:5. In the smaller part he plants 90 seedlings. How many similar seedlings could he plant in the bigger section?
4) A recipe for a cake requires 6 eggs. How many eggs would you need to make a cake 2½ times as big?
5) A cocktail requires 90ml of gin per 600ml of cranberry juice. If you fill a glass with 180ml of cranberry juice, how much gin should you add?
6) If the exchange rate from euros to pounds sterling is 1.20:1, how many euros would I get for £65?

Answers on page 174

When is a Number not a Number?

A nightmare to use if you're undertaking some complicated arithmetic, Roman numerals are nonetheless still reasonably commonplace, whether on inscriptions in churchyards or at the end of TV credits.

Quiz 17

Do you know what the following numerals represent?

1) I
2) M
3) C
4) L
5) V
6) X
7) D

Answers on page 174

All Latin to Me

There are a few rules about Roman numerals that are worth keeping in mind:

- *Roman numerals are arranged, as with our numbers, from the largest value to the smallest reading left to right, with each letter's value added to the previous one.*
- *Only I, X, C and M can be repeated in a row, but never more than three times.*

- *This means that certain numbers must be written using subtraction. That is to say, a letter with a smaller value precedes one with a larger value, with the smaller subtracted from the larger. For instance, since we cannot write IIII for 4, we use IV instead (effectively 1 from 5).*

Quiz 18

Can you solve the following sums (writing your answers in Roman numerals)?

1) II + V =
2) VIII – III =
3) XIII + XIX =
4) XLVI – XXIII =
5) CXXII + CXXVI =
6) CCCXII – CLXII =
7) CXLVIII + CCLII =
8) MDCCCXXII – CMXLV =

Answers on page 175

Binary Finery

The number system we use is decimal. That is to say, it uses the digits 0–9 and each column has a value to the power of 10 greater than the column to its right. In other words, we have a units column, then we move on to the tens column, then on to the hundreds, on further still to the thousands and so on.

In contrast, the binary number system uses only the digits 0 and 1 and the value of columns increases by the power of 2. So the first column in binary has a value of 1, the second column has a value of 2, the next column 4, the next column 8, the next column 16 and so on.

So to express the figure 1, we would write 1. But to express 2, we write 10 (that is 0 in the 1 column and 1 in the 2 column). To express 3 in binary, we write 11 (that is 1 in the 1 column + 1 in the 2 column). It might seem confusing but binary is the basis of our modern computer age!

Quiz 19

Look at the binary numbers below and express them as decimal numbers.

1) 100
2) 101
3) 111
4) 10000
5) 100011

Quiz 20

This time, rewrite the decimal numbers in binary.

1) 9
2) 11
3) 14
4) 25
5) 100

Answers on page 175

Quiz 21

Just as with decimals, binary numbers can have figures after a decimal point (or should that be binary point?) But rather than being worth tenths, hundredths, thousandths, etc., they are worth halves, quarters, eighths, and so on. In this quiz, express the binary numbers as decimal numbers.

1) 0.1
2) 0.11
3) 10.01
4) 1100.001
5) 100.111

Quiz 22

Finally, here are a few arithmetic questions using binary. Give your answers in binary too.

1) 1000 + 100 =
2) 10001 – 10 =
3) 111 x 111 =
4) 1000000 + 100000 – 110 =
5) 1.001 + 1.001 =
6) 101 x 11.001 =

Answers on page 176

Squared, Cubed and Diced

Indices are useful bits of shorthand used when a number is to be multiplied by itself a certain number of times. If 3 is to be multiplied by itself and n is an unspecified index (the singular version of indices), it may be written like this: 3^n. It might also be expressed as '3 to the power of n' or '3 to the nth power'.

But what does it mean in practice? Well, in the two most common indices that you are likely to encounter, n=2 or n=3. Using 3 as our base number again, 3^2 is usually expressed as '3 squared' (rather than '3 to the power of 2') and 3^3 is usually expressed as '3 cubed'.

3 squared means 3 is multiplied by itself once (3 x 3 = 9).

3 cubed means 3 is multiplied by itself twice (3 x 3 x 3= 27).

As you can probably guess, as the index number increases you soon find yourself dealing with very large numbers indeed. For instance, 3^9 (or 3 x 3 x 3 x 3 x 3 x 3 x 3 x 3 x 3) = 19,683.

Quiz 23

Here is a table that needs completing. Can you fill in the gaps?

Base number (A)	A^2	A^3
2	4	8
3	9	
4		64
5		
6		
7		
8		
9		
10		
11		
12		

Quiz 24

And what about these (calculators at the ready!)?

1) 4^4 =
2) 5^5 =
3) 6^6 =
4) 7^7 =
5) 8^8 =
6) 9^9 =
7) 10^{10} =

Answers on pages 176–7

To the Root of the Matter

While it is possible to square a number, it may also be 'unsquared'. To do this is to find the 'square root' of a number. So, for instance, we know that $4^2 = 16$. Thus the square root of 16 is 4. The symbol $\sqrt{}$ represents a square root (e.g. $\sqrt{16} = 4$). Similarly, there are cube roots ($\sqrt[3]{}$). We know that $4^3 = 64$, so the $\sqrt[3]{}$ of $64 = 4$.

Quiz 25

See how you get on with the following teasers:

1) $\sqrt{81}$
2) $\sqrt{169}$
3) $\sqrt[3]{125}$
4) $\sqrt{256}$
5) $\sqrt[3]{1000}$
6) $\sqrt[3]{1000000}$
7) $\sqrt{400}$
8) $\sqrt[3]{27000}$
9) $\sqrt{4900}$
10) $\sqrt[3]{2744}$

Answers on page 177

All in Order

Sometimes you might be confronted by a sum that has multiple elements to it. Rather like language, mathematics has some 'grammar rules' to help you know exactly what is being expressed and to avoid confusion.

Take the following sum: 5 x 3 + 4. If we do the addition bit first, we get 5 x 7 = 35. However, if we do the multiplication instruction first, we get a very different answer: 15 + 4 = 19.

Now, such imprecision is enough to drive your average maths boffin to the edge of despair. Therefore, there are some simple rules on the order in which to carry out instructions, which can be remembered by the handy mnemonic BIDMAS, starting with the instruction to be followed first down to the one to be followed last:

- *Brackets*
- *Indices*
- *Division*
- *Multiplication*
- *Addition*
- *Subtraction*

Going back to our original sum, according to these rules the multiplication should be done first. So, 5 x 3 + 4 = 19. If you want the addition done first, that part of the sum must go in brackets so that it reads 5 x (3 + 4) = 35.

Quiz 26

Have a look at the following five sums and decide where brackets should be added in each case:

1) $9 + 6 \div 2 = 12$

2) $7 - 3 + 8 = -4$

3) $112 \div 2 \times 7 + 2 = 394$

4) $5 \times 18 + 6 - 10 = 110$

5) $9 \times 7 \times 10 + 2 = 756$

6) $4^2 + 8 \times 15 \div 5 - 12 = 60$

Quiz 27

In these questions, follow the BIDMAS rules to come up with the right answers.

1) $9 + 5^2 - 6 =$
2) $7 - 6 \div 3 =$
3) $6 \times (6 + 5) + 7 =$
4) $4^3 + (9 - 2)^2 =$
5) $10 \times 8 \div 4 =$

Answers on pages 177–8

Next in Line

Numbers can have interesting patterns. Here, we consider how number sequences arise and how we work out further terms in a sequence.

Quiz 28

Work out the pattern that each sequence follows and then insert the next two numbers in the sequence.

1) 2, 4, 8, 16, 32, __, __
2) 2, 3, 5, 7, 11, __, __
3) 2, 5, 11, 23, 47, __, __
4) 3, 5, 8, 13, 21, __, __
5) 10, 100, 110, 1000, 1010, __, __
6) 2, 4, 10, 28, 82, __, __

Answers on page 178

Quiz 29: Arithmetic Crammer

1) What is 9 x 9?
2) What is –18 divided by –3?
3) If Farmer Giles has 186 sheep, Farmer Wendy had 297 sheep and Farmer Bob has 334 sheep, how many sheep do they have in total?
4) How many times does 18 go into 792?
5) What is the first prime number over 50?
6) If a recipe requires 350g of butter to create a meal that feeds four, how much butter do you need if the recipe is to feed 10?
7) What is the binary number 1000100 as a decimal?
8) What is the square root of 1,024?

Answers on page 179

A PART OF THE STORY: FRACTIONS AND PROBABILITY

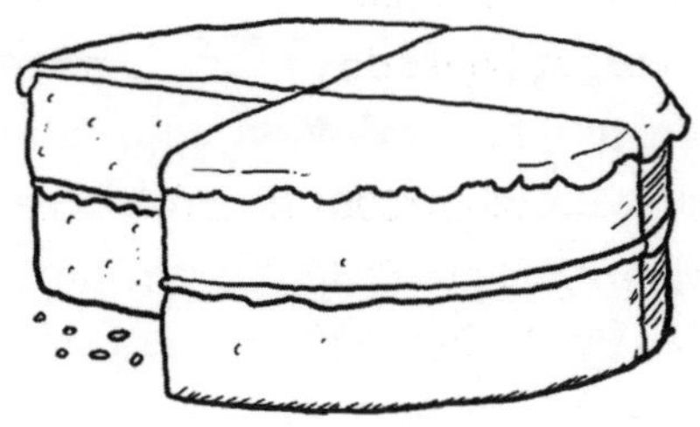

Maths would be a whole lot simpler if all the numbers it dealt with were whole. But life just isn't like that. Whether you're adding half a pound of flour to your recipe, walking a quarter of a mile up the road or spending two thirds of your life in the office, we are surrounded by fractions. A fraction is, after all, any smaller part of a whole number. For each number there are an infinite number of fractions.

A Little Vulgar

Let's start with vulgar fractions. These are the sort that are represented by one number sitting on a line above another. For instance, half as a vulgar fraction is ½. The value of a vulgar fraction is equal to the value of the top number (called the numerator) divided by the bottom number (the denominator).

Quiz 30

In this first quiz, rewrite the following values as vulgar fractions:

1) A quarter
2) three quarters
3) two thirds
4) five eighths
5) nine tenths

In Other Words...

There is a very useful concept known as the equivalence of fractions. This means that you can represent the same value in multiple ways, in a fashion quite impossible with whole numbers. For instance, all of the following vulgar fractions are equivalent to ½:

$^{2}/_{4}$ $^{3}/_{6}$ $^{4}/_{8}$ $^{5}/_{10}$ $^{16}/_{32}$ $^{79}/_{158}$

If you think of each fraction as a division sum (dividing the top number by the bottom number), you get the same value in each case.

Answers on page 179

Quiz 31

Rewrite the following fractions in their simplest form.

1) $^{48}/_{64}$
2) $^{90}/_{100}$
3) $^{80}/_{100}$
4) $^{37}/_{111}$
5) $^{5}/_{85}$
6) $^{420}/_{490}$

A Little Plus a Little

In order to be able to add or subtract fractions, equivalence of fractions is vital. This is because in order to carry out an addition or subtraction, the fractions must share a common denominator. We can add or subtract halves together, or sevenths together, but not a half and a seventh. So we go in search of the lowest common denominator (a far more useful thing in maths than it is when you're talking about popular culture, incidentally). The lowest common denominator is the lowest number that, in our case, 2 and 7 will divide into (the answer being 14).

The chief rule of equivalence is to do the same to the numerator as you do to the denominator. Let's say we were to add $^1/_2$ to $^3/_7$:

- *First, change both equations so that the denominator is 14 (our lowest common denominator).*

- *Starting with our half, we have to multiply the existing denominator (2) by 7 to get the lowest common denominator.*

Answers on page 179

So we must also multiply the numerator by 7. $^1/_2$ thus becomes $^7/_{14}$.

- *Now for $^3/_7$. The 7 is multiplied by 2 to get to 14, so the numerator is also multiplied by 2. $^3/_7$ thus becomes $^6/_{14}$.*

- *So now we are adding together $^7/_{14}$ and $^6/_{14}$. When adding or subtracting vulgar fractions, leave the lowest common denominator as it is and work your maths magic on the numerators only.*

- *We know 7 + 6 = 13 so our answer is $^{13}/_{14}$.*

- *If we had wanted to subtract $^3/_7$ from $^1/_2$, our sum would have been $^7/_{14} - {}^6/_{14} = {}^1/_{14}$.*

Quiz 32

1) $^5/_9 - {}^3/_9 =$
2) $^3/_5 + {}^3/_5 =$
3) $^3/_5 + {}^3/_{10} =$
4) $^3/_4 - {}^2/_3 =$
5) $^2/_7 + {}^7/_9 =$
6) $^5/_8 - {}^4/_7 =$

Answers on page 180

Part-Timers

The rules for multiplying fractions are, needless to say, different again. However, it might surprise you to learn that it's actually easier than adding or subtracting them because you don't need to do any equivalence conversions. You simply multiply the numerators together and then do the same with the denominators. So $^2/_3 \times {}^3/_4 = {}^6/_{12}$. If you want to you can now do an equivalence conversion on the result: $^6/_{12}$ is simply another way of expressing $^1/_2$.

And remember, if you're multiplying a fraction by a whole number, the whole number can be expressed as $\frac{n}{1}$ (e.g. if your number n is 7, it can be written as $^7/_1$ if it helps you to do your calculation).

Quiz 33

Give your answers to the following questions by putting the resulting fractions into their simplest form.

1) $2 \times {}^4/_9 =$
2) $4 \times {}^2/_5 =$
3) $2 \times {}^7/_8 =$
4) $3 \times {}^8/_{11} =$
5) $3 \times {}^8/_{10} =$
6) $6 \times {}^9/_{14} =$

Answers on page 180

Mixing It Up

A mixed number is one that includes a whole number and a fraction, such as $1\frac{1}{2}$. Adding or subtracting them poses few problems since you add or subtract the whole numbers and then the fractions separately.

However, you can't treat mixed numbers like that when multiplying them or you'll get the wrong answer. Therefore, you need to convert mixed numbers into top-heavy fractions (also known as improper fractions). To do this, use the denominator in the fraction part of the mixed number as the denominator in your top-heavy fraction too. So in $1\frac{1}{2}$, 2 is the denominator. 1 may be expressed as $\frac{2}{2}$ so to get the top-heavy fraction we add $\frac{2}{2} + \frac{1}{2}$ and get $\frac{3}{2}$.

Quiz 34

Convert the following mixed numbers into top-heavy fractions, and top-heavy fractions into mixed numbers:

1) $1\,{}^{3}/_{5}$

2) $3\,{}^{3}/_{4}$

3) $2\,{}^{2}/_{5}$

4) $7\,{}^{5}/_{8}$

5) $4\,{}^{71}/_{100}$

6) ${}^{9}/_{8}$

7) ${}^{35}/_{16}$

8) ${}^{53}/_{7}$

9) ${}^{143}/_{12}$

10) ${}^{57}/_{17}$

Answers on page 180

Quiz 35

In this quiz, mark whether each statement is true or false.

1) $^{28}/_{5} = 5\ ^{3}/_{5}$
2) $^{78}/_{5} < {}^{154}/_{5}$
3) $^{10}/_{9} < 1\ ^{2}/_{9}$
4) $^{20}/_{11} = {}^{58}/_{33}$
5) $17\ ^{3}/_{8} < {}^{150}/_{9}$
6) $^{42}/_{5} > {}^{50}/_{6}$

Grief Cancelling

When you are multiplying fractions, it is sometimes possible to make a particularly grim-looking sum look a little less unfriendly. This is because it does not matter which numerator goes on which denominator, so allowing for a process called cancelling down (which relies once again on equivalence of fractions).

Here's how it works. Say you have a sum of $^{11}/_{14}$ x $^{7}/_{13}$. Doing the 11 x 7 bit is no problem at all but 14 x 13 is rather more of a challenge. We know it's OK to swap the denominators over, so we are left with a new sum of $^{7}/_{14}$ x $^{11}/_{13}$. Now, we can cancel down $^{7}/_{14}$ to $^{1}/_{2}$. So the new sum is $^{1}/_{2}$ x $^{11}/_{13}$. After a simple calculation we thus reach an answer of $^{11}/_{26}$.

Answers on page 181

Quiz 36

Solve the following sums by cancelling down.

1) $^{2}/_{19} \times {}^{5}/_{6} =$
2) $^{3}/_{7} \times {}^{29}/_{33} =$
3) $^{6}/_{7} \times {}^{53}/_{54} =$
4) $^{4}/_{5} \times {}^{15}/_{16} =$
5) $^{18}/_{19} \times {}^{11}/_{27} =$
6) $^{27}/_{32} \times {}^{42}/_{81} =$

A Little of a Little

Rather surprisingly, the trick to dividing fractions is to multiply them! But before you multiply them, you must turn the second fraction in your sum (the one you're dividing by) upside down – a process known as inverting. Take this example:

$^{2}/_{7} \div {}^{2}/_{5}$ becomes $^{2}/_{7} \times {}^{5}/_{2}$, which equals $^{10}/_{14}$.

Answers on page 181

Quiz 37

Work out the following sums and remember, if your answer is a top-heavy fraction, convert it into a mixed number.

1) $\frac{1}{2} \div \frac{3}{4} =$

2) $\frac{2}{3} \div \frac{7}{8} =$

3) $\frac{1}{8} \div \frac{1}{8} =$

4) $\frac{5}{6} \div \frac{3}{7} =$

5) $\frac{7}{9} \div \frac{1}{2} =$

6) $\frac{11}{14} \div \frac{9}{10} =$

Answers on page 181

To the Point

Another way of expressing a part of a number is by using decimal points. As discussed on page 54, each column after a decimal point has a steadily decreasing place value. So 0.5 (equivalent to $\frac{1}{2}$) is bigger than 0.05 (equivalent to $\frac{1}{20}$).

The good news is that when it comes to adding or subtracting decimals, the rules are exactly the same as when dealing with whole numbers. The only thing to remember is that if you have different numbers of columns, it's worth putting in a 0 to equalize the columns. So:

9.82 –	*becomes*	9.820 –
6.457		6.457

Quiz 38

Keeping that rule in mind, have a go at the following sums.

1) 1.8 + 3.6 =
2) 7.2 – 4.5 =
3) 1.875 + 0.42 =
4) 6.843 – 3.92 =
5) 7.5 + 2.87 + 3.623 =
6) 9.842 – 6.95 =

Point It Out

When you multiply with decimals, it is often a help to remove the decimal point while you're doing the calculation and put it back in once you have your result. But where should it go back in, you ask? The trick is to end up with an answer that has the same number of decimal places as there were in the original sum. Take these two examples:

10 x 0.5 *Remove the decimal point to get 10 x 5*

10 x 5 = 50 *You had one decimal place in the original sum so add one place to the answer*

5.0 *So your answer is 5!*

Answers on pages 182–3

5.3 x 6.9	*Remove the decimal points to get 53 x 69*
53 x 69 = 3657	*You had two decimal places in the original sum so add two places to the answer*
36.57	*So your answer is 36.57!*

Remember, if both decimals are less than 1, you quickly end up with very small answers!

Quiz 39

Now try these ones yourself:

1) 12 x 0.7 =
2) 6 x 1.23 =
3) 0.7 x 2.2 =
4) 0.5 x 3.3 =
5) 1.8 x 1.8 =
6) 4.6 x 2.1 =

Answers on page 183

Cut to the Point

If you're dividing a decimal by a whole number, things are reasonably straightforward. You treat it like a normal division sum, only remembering to insert a decimal point at the correct point:

e.g. 19.6 ÷ 4 becomes

$$\begin{array}{r|l} & 0\,4\,.\,9 \\ \hline 4 & 1\,{}^{1}9\,.\,{}^{3}6 \end{array} \quad \text{so } 19.6 \div 4 = 4.9$$

However, it's a whole different ball game if you're dividing a decimal by another decimal. We turn, perhaps surprisingly, to equivalence of fractions again.

The trick is to think of your sum as a vulgar fraction. For instance, 23.8 ÷ 0.4 can also be expressed as $^{23.8}/_{0.4}$. Now, as we have seen, it is a lot easier to deal with whole numbers, so we turn the denominator into one. A bit of simple arithmetic reveals that if we multiply 0.4 by 5 we get 2. And if we are scaling up the denominator, we must do the same to the numerator, so 23.8 also gets multiplied by 5. Our new sum is $^{119}/_{2}$. Much simpler figures to deal with:

$$\begin{array}{r|l} & 0\,5\,9\,.\,5 \\ \hline 2 & 1\,{}^{1}1\,{}^{1}9\,.\,{}^{1}0 \end{array}$$

So 119 ÷ 2 = 59.5. And so does 23.8 ÷ 0.4. Magic!

Quiz 40

Your turn to put theory into practice:

1) 6.4 ÷ 10 =
2) 11.2 ÷ 8 =
3) 16.8 ÷ 7 =
4) 4.5 ÷ 1.5 =
5) 18.2 ÷ 0.8 =
6) 3.4 ÷ 2.7 =

Preaching to the Converted

Depending on the circumstances and on an individual's preference, it is useful to be able to express a fraction as a vulgar fraction, a decimal fraction or a percentage.

Many of these equivalent terms are so familiar that you'll know them off by heart but there are some rules to help with any that don't roll off the tongue.

- *To convert a decimal into a percentage, multiply the decimal by 100 and add a percentage symbol at the end (e.g. 0.2 ⇨ 0.2 x 100 = 20%).*

- *To convert a percentage into a decimal, remove the percentage sign and divide the figure by 100 (e.g. 20% ⇨ 20 ÷ 100 = 0.2).*

- *To convert a vulgar fraction into a percentage, multiply the fraction by $^{100}/_{1}$ and add a percentage symbol at the end (e.g. $^{1}/_{5}$ ⇨ $^{1}/_{5}$ x $^{100}/_{1}$ = $^{100}/_{5}$ = 20 = 20%).*

Answers on page 183

- *To convert a percentage into a vulgar fraction, remove the percentage sign and divide the figure by 100 (e.g. 20% ⇨ $^{20}/_{100} = {}^{1}/_{5}$).*

- *To convert a vulgar fraction into a decimal, treat your vulgar fraction as a simple division.* $^{1}/_{5}$ ⇨

$$\begin{array}{r|l} & 0\;.\;2 \\ \hline 5 & 1\;.\;{}^{1}0 \end{array}$$

- *To convert a decimal into a vulgar fraction, you need to think in terms of place value. In 0.2, 2 is in the tenths column. So do the following sum: $0.2 = {}^{2}/_{10} = {}^{1}/_{5}$. If you have multiple figures after the decimal it gets trickier. For instance, $0.25 = {}^{2}/_{10} + {}^{5}/_{100} = {}^{20}/_{100} + {}^{5}/_{100} = {}^{25}/_{100} = {}^{1}/_{4}$. If there was a figure in the thousandths column, you're lowest common denominator would be 1,000, and so on for every additional figure after the decimal point. Remember always to cancel down your final result too.*

Quiz 41

Fill in the gaps in the table below.

Vulgar	Decimal	Percentage
$\frac{1}{50}$	0.02	2%
$\frac{1}{40}$	____	____
____	0.04	____
____	____	5%
____	____	10%
____	0.125	____
$\frac{1}{5}$	____	____
____	____	25%
____	$0.\dot{3}$	____
____	0.375	____
$\frac{2}{5}$	____	____
____	____	50%
____	0.6	____
____	____	62.5%
____	$0.\dot{6}$	____
$\frac{7}{10}$	____	____
____	____	75%
$\frac{4}{5}$	____	____
____	0.875	____
____	0.9	____
____	____	100%

Answers on page 184

The Bargain Hunter's Guide to Percentages

Per cent derives from the Latin *per centum*, meaning per 100. So when you use percentages, you are expressing a number as a fraction of 100. If we talk about 5%, it could also be expressed as $^{5}/_{100}$; 70% = $^{70}/_{100}$ and so on.

Using percentages often makes things clearer than they might be if expressed by a vulgar fraction or a decimal. For instance, if a tax rate increases from 15% to 17.5%, that is much easier to understand than saying it has increased from three-twentieths to seven-fortieths.

In most cases, when calculating percentages it is normally in terms of a 'percentage of something else'. For instance, a sign in a shop that tells you a particular jacket is for sale at 15% off ('the percentage') is meaningless if we don't know the price ('the something else'). If the jacket costs £30, we can work out what 15% is with a bit of simple arithmetic:

$$^{15}/_{100} \times {}^{30}/_{1} = {}^{450}/_{100} = 4.5$$

So, 15% off equals £4.50 off the price tag. Your £30 jacket is now available for £25.50.

Quiz 42

Here are five more sale items. In each case, how much is the reduction and what is the new price?

1) A pair of trousers that originally cost £40 is on sale at 10% off.
   Value of reduction _____ New Price _____
2) A computer that originally cost £560 is on sale at 25% off.
   Value of reduction _____ New Price _____
3) A pair of shoes that originally cost £120 is on sale at 18% off.
   Value of reduction _____ New Price _____
4) A dinner service with minor imperfections originally cost £750 but has been reduced by 60%.
   Value of reduction _____ New Price _____
5) A car that originally cost £10,250 is on sale at 20% off.
   Value of reduction _____ New Price _____

Answers on page 185

Highly Rated

Sometimes you may have some raw figures that you want to convert into an easily graspable percentage rate. For instance, if 13 out of 25 pupils pass an exam, what percentage is the pass rate? The key is to turn your two known figures into a vulgar fraction (in this case $^{13}/_{25}$) and multiply both by 100. Try to cancel down the denominator and 100 wherever possible. So the sum becomes $^{13}/_{25}$ x 100 = $^{13}/_{1}$ x 4 = 52. So the pass rate is 52%.

Quiz 43

1) In a survey, 8 out of 10 people reported that their favourite type of chocolate is milk chocolate. Express this as a percentage.
2) Of 20 people asked in a survey, 7 said they had eaten Thai food within the last week. What percentage of respondents had recently eaten Thai food?
3) 78 out of 120 local residents had seen foxes in their gardens. What percentage of residents reported seeing the animals?
4) Of 160 people surveyed, only 28 people had visited India. What percentage of respondents had been to India?
5) 800 people were polled about where they holidayed. 504 said they always went abroad. What percentage of people in the survey holiday abroad?

Answers on page 185

Ups and Downs

Two of the most common but confusing calculations you might need to make when working with percentages are percentage increases and percentage decreases. Here's how to do them:

- *Calculating a percentage increase. Your favourite sandwich goes up in price from Value A (£2.00) to Value B (£2.12). Firstly, subtract A from B (2.12 – 2.00 = 0.12). Divide this result by A (0.12 ÷ 2 = 0.06) and multiply the result by 100 (0.06 x 100 = 6). The sandwich has risen in price by 6%.*

- *Calculating a percentage decrease. If the same legendary sandwich fell in price from Value A (£2.00) to Value B (£1.75), to work out the percentage decrease, take Value B away from A (2.00 – 1.75 = 0.25). Divide the result by Value A (0.25 ÷ 2 = 0.0125) and multiply that figure by 100 (0.125 x 100 = 12.5). The price has decreased by 12.5%. Best be quick and buy that sandwich before they all sell out.*

Quiz 44

Once you've had a go at the following questions, see what percentage you got correct!

1) A guest house charged £40 per night but increased their prices to £65. What is the percentage increase in price?
2) Meanwhile, at the hotel bar, a bottle of red wine fell in price from £12.20 to £10.90. What is the percentage decrease in price?
3) The number of families staying each year rose from 760 to 830? What is the percentage increase?
4) The number of honeymoon couples fell from 67 to 54. What is the percentage decrease?
5) The number of business delegates rose from 1,213 to 1,347. What is the percentage increase?

Answers on page 185

Mission (Im)Possible: Probability

It was Benjamin Franklin who said that 'in this world nothing can be said to be certain, except death and taxes'. Indeed, we all spend a great deal of our lives calculating the likelihood of particular outcomes and using our (largely non-mathematical) conclusions to guide our decisions. Mathematicians, though, demand a little more rigorous logic in figuring out probability. At a basic level, any given event may be classified as possible or impossible.

Quiz 45

Decide which of the following events are mathematically impossible and which are possible. But remember, all sorts of events that are extremely unlikely are not actually impossible. For instance, it is improbable that we will be able to time travel next week but it is not out of the question!

1) Win the lottery.
2) Roll a 7 on a standard dice.
3) Deal five aces from a standard pack of playing cards.
4) Visit Mars.
5) Own the *Mona Lisa*.
6) Pick a blue ball from a bag full of red balls.

Answers on page 186

Definitely Maybe

We might also think of events in terms of being certain or uncertain. But just as we may be surprised at how few things are mathematically impossible, there are probably fewer 'dead certs' than you'd expect too.

Quiz 46

Read the following statements and consider which are irrefutable (and therefore certainties) and which are uncertainties (despite the fact that we might be very surprised if they didn't happen!).

1) The sun will rise tomorrow.
2) If I roll a standard dice, I will roll a number between 1 and 6.
3) No human will run 100m in under 5 seconds.
4) I will be older tomorrow than I am today.
5) My football team will either win, lose, or draw the next match they play.
6) Cats will always hunt mice.

Answers on page 186

Toss a Coin for It

If that's all getting too philosophical, let's take a look at some more traditional probability questions.

We all know that if you flip a coin, there is a $\frac{1}{2}$ chance that it will land on heads and a $\frac{1}{2}$ chance that it will be tails. And you have a $\frac{1}{6}$ chance of rolling any given number on a standard six-sided dice or a $\frac{1}{52}$ chance of picking a particular card from a standard playing deck.

But how do we calculate the probability of more complex outcomes? Here is where our knowledge of multiplying fractions comes in useful. For example, if we wanted to find out the probability of a coin landing on tails twice in a row, we know that with each toss there is a $\frac{1}{2}$ chance, so we must multiply $\frac{1}{2}$ by $\frac{1}{2}$. The probability of a coin landing on tails twice consecutively is therefore $\frac{1}{4}$.

Quiz 47

What are the odds of the following events?

1) Getting two heads from two consecutive coin flips.
2) Getting three tails from three consecutive coin flips.
3) Flipping five consecutive heads.
4) Rolling 1 or 2 on a standard six-sided dice.
5) Rolling 1 or 2 on a standard six-sided dice twice consecutively.
6) Rolling 6 on a standard six-sided dice twice consecutively.
7) Rolling 6 on a standard six-sided dice six times consecutively.
8) Picking the Ace of Hearts from a playing deck at random.
9) Picking a diamond.
10) Picking a picture card.
11) Picking two red picture cards consecutively.
12) Picking the four aces consecutively.

Answers on page 187

Could it be you?

In this quiz we are going to discover just how likely it is that you will be able to retire on the proceeds of winning the lottery or achieving the best possible hand in poker.

Quiz 48

1) In order to win the lottery, you need to correctly match six numbers randomly selected from all the numbers between 1 and 49. The numbers can appear in any order and any number can be selected only once. Can you calculate your odds of success?
2) In poker, the supreme hand is a royal flush, consisting of the 10, Jack, Queen, King and Ace of a single suit. Assuming you're playing a poker variant in which any hand must be created from 5 randomly dealt cards from a standard pack of 52 playing cards, what are your chances of sweeping aside all opposition?

Answers on page 187

Quiz 49: Fractions and Probability Crammer

1) What is 85% expressed as a vulgar fraction in its simplest form?
2) What is $^{13}/_{18} - ^{2}/_{27}$?
3) What is 13.8 divided by 0.6?
4) What do you get if you multiply $^{9}/_{13}$ by $^{17}/_{18}$?
5) What is the result if you divide $^{6}/_{7}$ by $^{3}/_{8}$?
6) What is 35% of £70?
7) You bought a jacket for £135 last year, when it was at the height of fashion. This year it is on sale for £80. What is the percentage decrease in price to the nearest per cent?
8) You roll a 12-sided dice twice consecutively. What are the odds of rolling a 4 or less and an 8 in any order?

Answers on page 188

MAKING SENSE OF THE DATA: STATISTICS

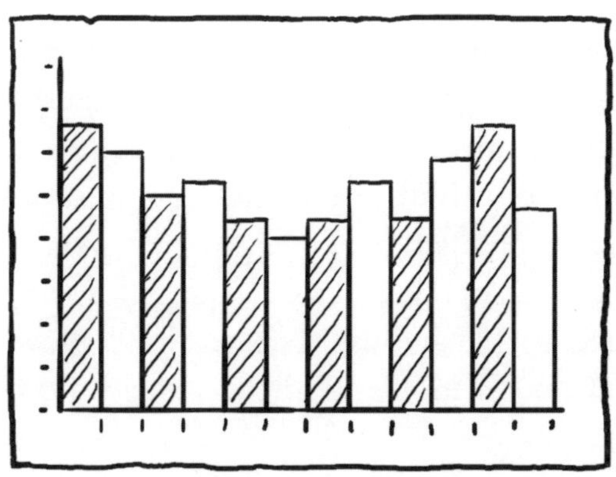

Data is raw information – facts and figures for us to make sense of. Data (the plural of datum, incidentally) might be seen as bricks that on their own are of little use, but which when brought together by someone with skill can be turned into a beautiful and useful building.

The area of mathematics known as statistics aims to make sense of data. This is, of course, easier said than done and is reliant on human interpretation. Ever heard the phrase, 'Lies, damned lies and statistics'? We are often nervous of statistics because it can be difficult to understand exactly how certain data has been manipulated.

But cast the cynic within you aside for now. No one is trying to pull the wool over anyone's eyes here! Getting to grips with some basics of statistics will provide you with great tools for life.

Statistics offer a way to build up a big picture from a small pool of information. Think of a marketing company. If they want to test out a new product, they don't contact every person in the country to ask their opinion. Instead, they ask a sample group of, say, a hundred people and use their results to draw conclusions about what the population as a whole is likely to think.

Quantity, Not Quality

There are two main types of data: Qualitative and Quantitative. In short, quantitative data can be expressed in figures and qualitative cannot.

Quiz 50

Have a look at the follow questions from a survey and decide if they are dealing with qualitative or quantitative data.

1) How many people are resident in your household?
   Qualitative ___ Quantitative ___
2) How many residents are under 18 years old?
   Qualitative ___ Quantitative ___
3) Which football team do you support?
   Qualitative ___ Quantitative ___
4) How many cars does your household own?
   Qualitative ___ Quantitative ___
5) What is your favourite brand of car?
   Qualitative ___ Quantitative ___
6) What colour is your car?
   Qualitative ___ Quantitative ___

Answers on page 188

Pick a Number

As this is a maths book, we shall restrict ourselves to quantitative data here. It comes in two major types too: discrete or continuous. Discrete data is recognizable because it can only be expressed by whole numbers. Continuous data can have any value, getting more detailed the more precise the method of measurement that is used.

Quiz 51

Decide which of the following questions would throw up discrete data and which would provide continuous data.

1) How tall are you?
   Discrete ___ Continuous ___
2) How many children do you have?
   Discrete ___ Continuous ___
3) How many properties do you own?
   Discrete ___ Continuous ___
4) How far do you commute to work?
   Discrete ___ Continuous ___
5) What size feet do you have?
   Discrete ___ Continuous ___
6) At what time do you wake up in the morning?
   Discrete ___ Continuous ___

Answers on page 189

Mr Average

One of the best ways to make some sense of a broad range of data is to find an average. If you have a class of 21 children, you might have trouble memorizing all 21 of their test scores but you could probably bring to mind their average score more easily.

That said, the word 'average' is bandied about very freely these days and there are always nuances. In mathematics, when we talk of averages, it is worth establishing whether you mean the mean, median or mode.

- *The mean is the most familiar form of average. Indeed, when people talk of averages, they often mean 'mean'! It is calculated by adding together the values of each datum in a data set and then dividing the result by how many data there are. So, to find the mean test score of our class of 21, add together the total of their scores and divide by 21. An oddity of the mean is that it might give a result that does not actually correspond directly to any single one of the data. That is to say, the mean score of our sample group might be 63.9% (let's call it 64%) but not one of the class actually scored that mark precisely. Therefore, we sometimes use...*

- *The median. This is the middle datum in a data set. In a class of 21, it's the 11th datum. In our data sample, it's: ~~24, 31, 35, 38, 46, 54, 57, 59, 62, 65~~, 66, ~~67, 67, 68, 70, 72, 74, 77, 77, 82, 87~~. So 66 is the median average and represents a score that one pupil actually got. Where you have an even number of data, find the median by taking the mean of the middle two numbers.*

- *Finally, there is the mode. This is the result that occurs most frequently in a data set. It is possible to have more than one mode. Indeed, that is the case with our class. Two pupils scored 67% and two scored 77%. So the modes in this case are 67 and 77. If the child who scored 66 had got an extra mark, the mode would be 67, as there would have been three pupils with that score and only two with 77%.*

Quiz 52

Below are five data sets. In each case, work out the mean, median and mode.

1) Height in cm of 5 respondents:
   155, 167, 176, 189, 210
   Mean = ____ Median = ____ Mode = ____
2) Hourly earnings in £ of 9 workers:
   6.70, 6.95, 7.20, 7.45, 7.45, 8.20, 8.60, 9.80, 11.40
   Mean = ____ Median = ____ Mode = ____
3) Age in whole years of 14 respondents:
   14, 18, 20, 20, 23, 27, 31, 35, 39, 43, 49, 56, 65, 74
   Mean = ____ Median = ____ Mode = ____
4) Length in minutes of 11 movies:
   97, 103, 110, 114, 116, 116, 119, 122, 124, 125, 129
   Mean = ____ Median = ____ Mode = ____
5) Exam results in per cent of 18 students:
   11, 24, 29, 33, 38, 43, 45, 45, 48, 51, 53, 53, 53, 56, 58, 60, 62, 65
   Mean = ____ Median = ____ Mode = ____

Affairs of the Chart

Filling pages with long strings of numbers can be problematic. So it is worth thinking about alternative mean of presenting data. This is where we enter the realm of the chart and table.

Answers on page 189

A Man Walks into a Bar Chart

Bar charts are great for expressing data so that you can quickly see relative proportions.

Quiz 53

In the example below, 100 people were asked what their favourite sport is. What can you learn from the table?

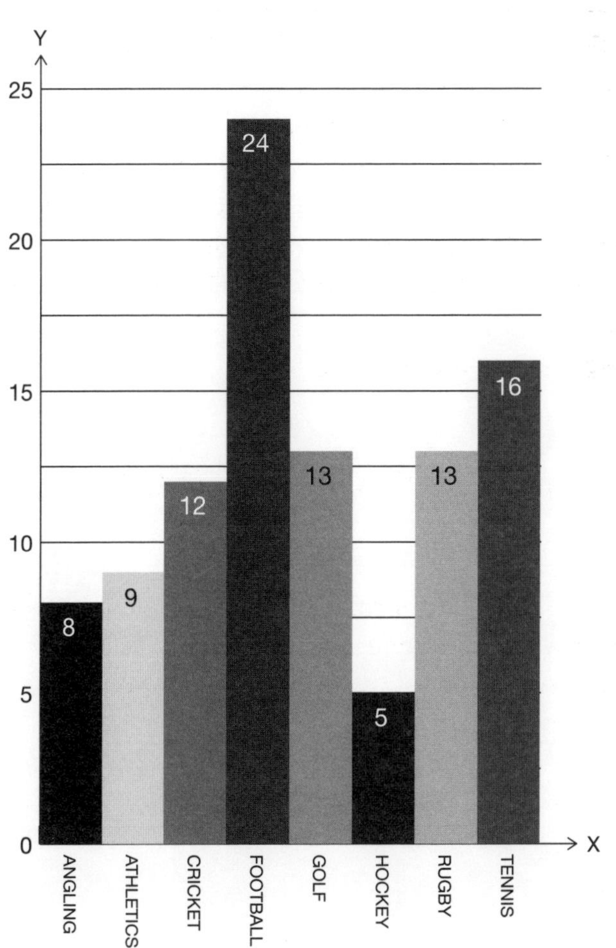

Answers on page 190

1) Which is the most popular sport?

2) Which is the second least popular?

3) How many people in total voted for either rugby or hockey?

4) How many people opted for sports where the only specialist equipment needed is a ball?

5) How many people chose sports that don't involve a ball?

Answers on page 190

Pie in the Sky

Another way to present proportions in an eye-catching way is through the pie chart. As you can see from the example below, a pie chart is so called because it looks very much like a pie cut into slices, the size of each slice corresponding directly to the data it represents.

Quiz 54

In this chart, Barney has noted down the various colours of 40 cars that passed his house.

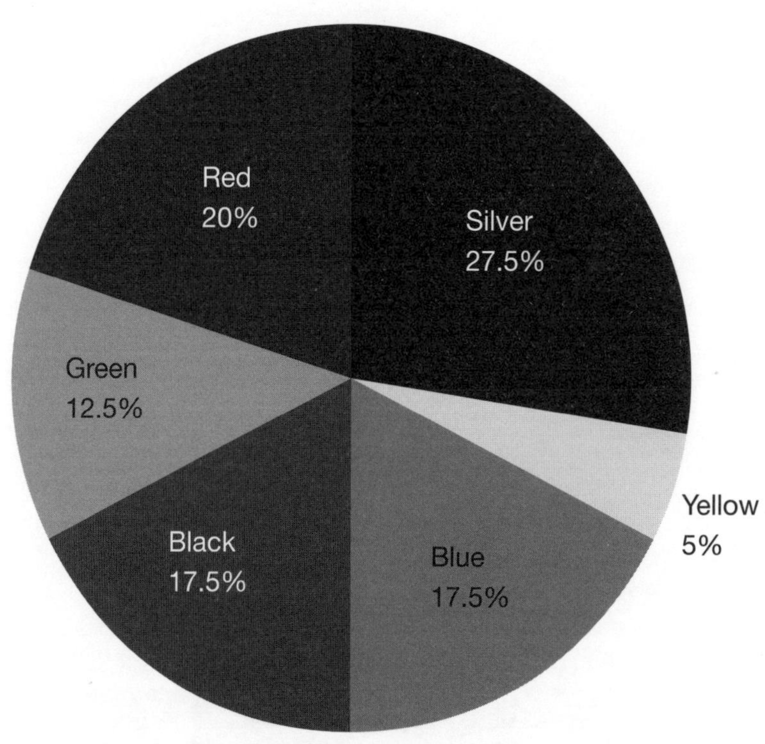

Answers on page 190

Now, answer the following questions:

1) How many degrees of the circle represent each car?
2) Which colour car was seen least often?
3) How many silver cars were there?
4) Which two colours were equally common and how many of each was seen?
5) How many green or red cars were noted?

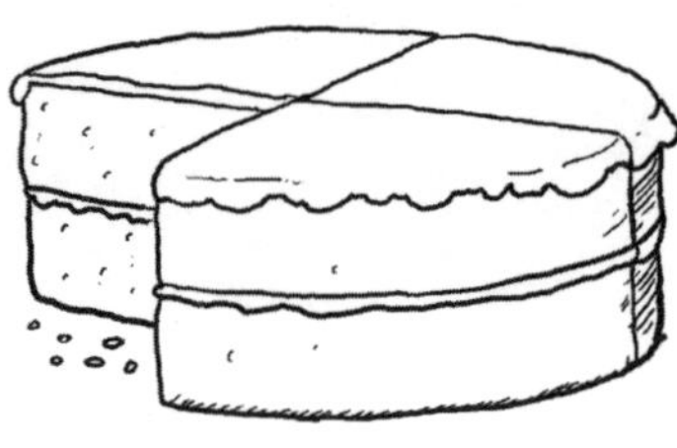

Answers on page 190

Uncharted Territory

In this exercise, the data you are using comes from the records of a birdwatcher who noted the different species he saw over a three-day period.

Quiz 55

Represent the data first in a pie chart and then in a bar chart. You'll need a protractor and a ruler for this.

Data

Species	*Number seen*
Blackbird	7
Blue tit	6
Crow	4
Kingfisher	1
Magpie	3
Sparrow	9

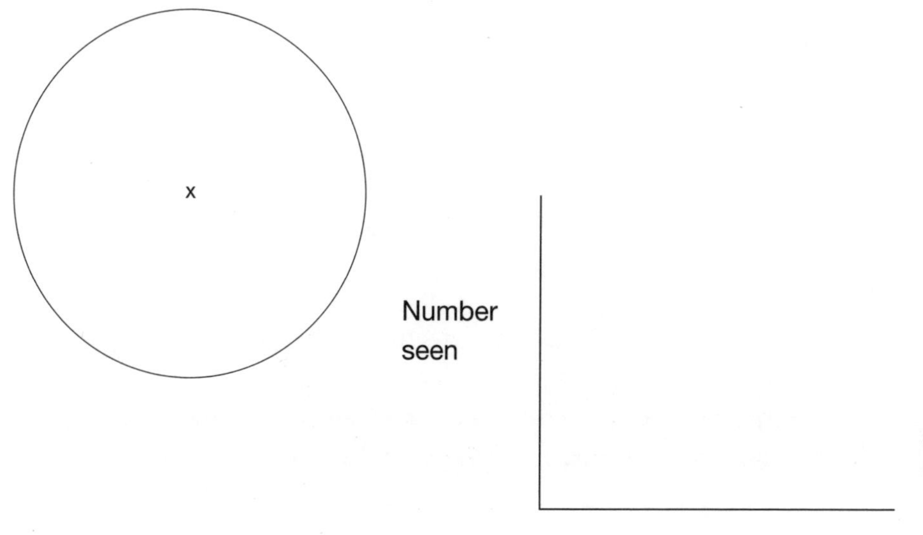

Answers on page 191

Scatterbrained

While pie and bar charts are great for illustrating proportions within a single data set, you will often have two data sets that are interrelated. In these instances, you might want to illustrate your data using a scattergraph.

Here's how they work. Let's say that a scientist has studied a sample of plants to see how high they grow in relation to their age. On a scattergraph he has a scale along the x-axis (let's say 'age' in this instance) and another on the y-axis ('height'). He marks as many points on the graph as there are sample plants. Each show where a plant's age on the x-axis intersects with its height on the y-axis.

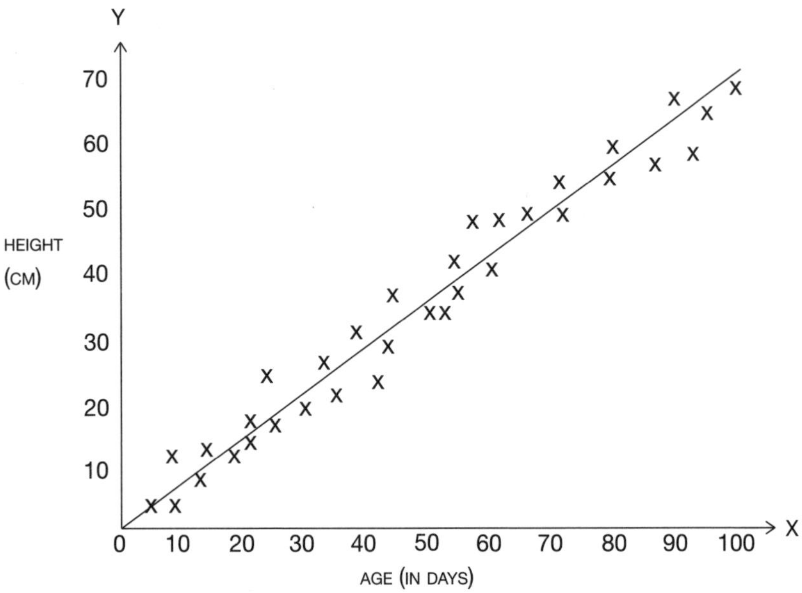

As you can see, a broad pattern emerges, with the points going from the bottom left of the graph to the upper right. This is why scattergraphs are so useful – they can show trends and thus help you make predictions. Looking at the graph above, you could conclude that plants get taller as they get older and that the older they get the taller they will become (though presumably they will get to a point at which they no longer grow).

The pattern evident in this table is known as a positive correlation. In a negative correlation, the points will go broadly from top left to bottom right. And sometimes there will be no correlation at all. For instance, if you took a sample group of ten adults and plotted their height against earnings you might not expect to see any correlation.

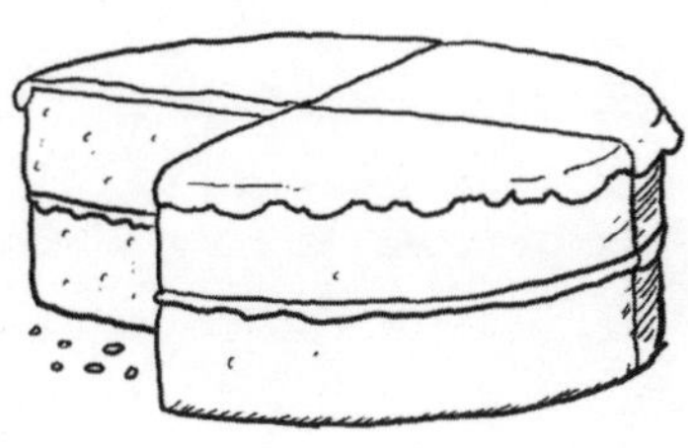

Quiz 56

Have a look at the descriptions below of several scattergraphs and decide whether you would expect it to show a positive correlation, a negative correlation, or no correlation at all.

1) Age of child (up to 12 years old) against time taken to run 100 metres.
   Positive correlation ____ Negative correlation ____ No correlation ____
2) Number of holidays taken per year against height.
   Positive correlation ____ Negative correlation ____ No correlation ____
3) Wealth against size of house.
   Positive correlation ____ Negative correlation ____ No correlation ____
4) Shoe size against number of friends on Facebook.
   Positive correlation ____ Negative correlation ____ No correlation ____
5) Cigarettes smoked against incidence of lung cancer.
   Positive correlation ____ Negative correlation ____ No correlation ____
6) Length of commute against time spent with family.
   Positive correlation ____ Negative correlation ____ No correlation ____

Answers on page 192

Nothing Venn-tured, Nothing Gained

The Venn diagram is a different kind of way to express data altogether and is part of a branch of mathematics called Set Theory. The Venn diagram was the creation of John Venn, a Victorian philosopher who devised it to show how different sets (they might be numbers or objects or even ideas) are interrelated.

- *A Venn diagram starts with a 'universal set'. This is essentially all the terms of reference in your investigation. For our purposes, let's look at a group of 10 school-age children. We draw a rectangle and inside it put the names of the 10 children.*

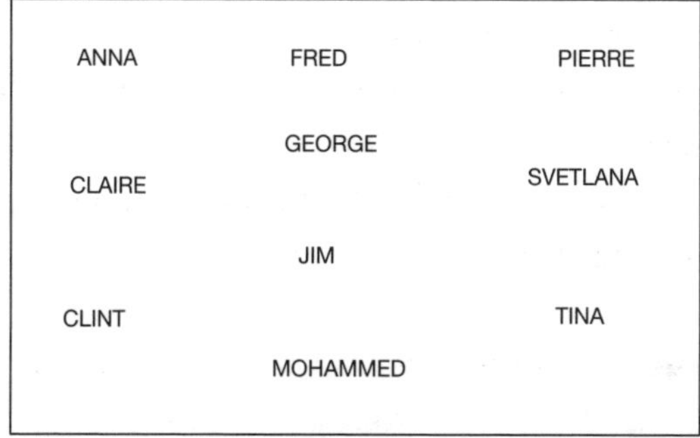

- *Now we can start introducing our subsets, with each subset going into its own circle. Let's start by dividing our universal set into girls (Set A) and boys (Set B).*

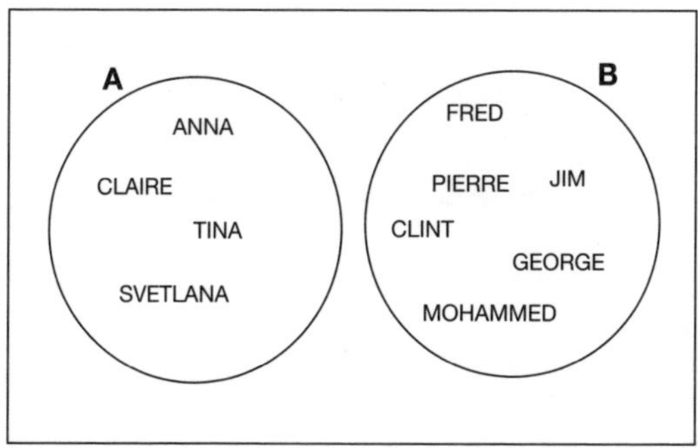

- *Next, let's have a subset of children who play football (Set C). This group comprises Claire, Clint, Fred, George, Mohammed, Pierre and Tina.*

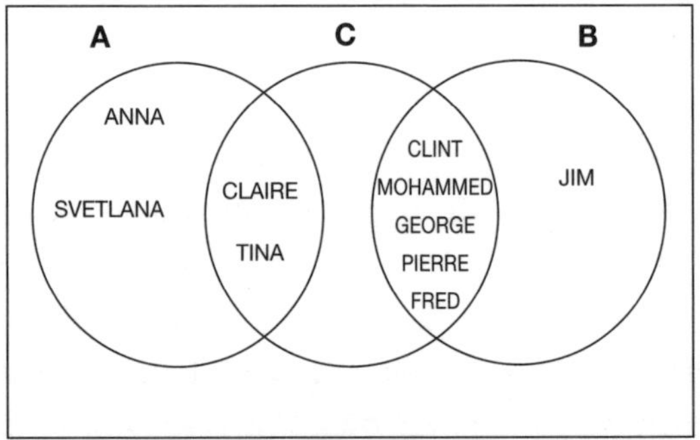

Quiz 57

Below you have a universal set of numbers. Add the data to the Venn diagram depicted below and composed of the three following subsets: Multiples of 3; Odd numbers; Multiples of 5.

Universal Set

2, 5, 6, 7, 9, 10, 12, 13, 15, 16, 20, 21 and 30

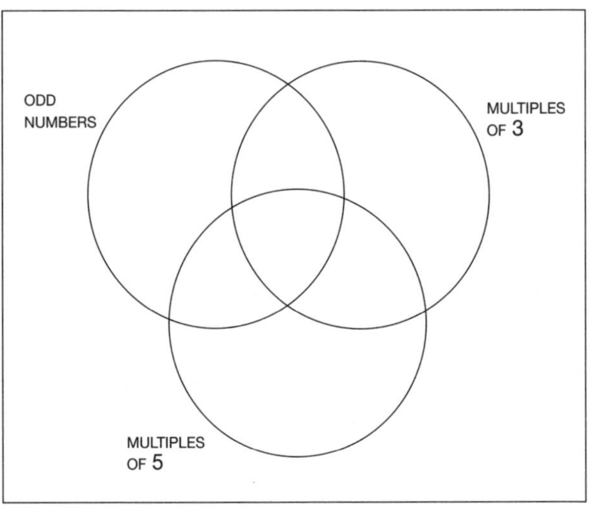

Answers on page 192

Unit-ed We Stand

For most of us mortals (or Non-Pure Mathematicians, as we are also known), numbers make much more sense when they have a bit of context.

Consider the following questions. How tall are you? How long did it take you to read this? If someone answered, respectively, 200 and about 7, we are not much the wiser. But if they said 200cm and about 7 seconds, we are in business.

Thankfully there is an organization that establishes internationally recognized units of measurement for pretty much anything you can think of. These are known as SI units (or Système International d'Unités). They generally work on the metric system too and the most basic units are:

kilogram (kg) for mass

second (s) for time

metre (m) for length

Little and Large

Sometimes we need to talk about things that are either very big or absolutely tiny. Luckily, there are a host of prefixes that help us to do this. For instance, we talk about millimetres, with the 'milli' telling us we are dealing in thousands of a metre. This means that we can talk about something being 2mm long rather than being 0.002m.

Quiz 58

The table below lists some of the most useful prefixes (especially in our age of computers and microchips). Can you complete the table by putting in the value of each multiplier?

Prefix (+ abbreviation)	***Multiplier***
Milli (m)	1 thousandth
Mega (M)	
Hecto (h)	
Nano (n)	
Giga (G)	
Centi (c)	
Tera (T)	
Micro (μ)	
Kilo (k)	

Answers on page 193

Putting Your Foot in It

Of course, while the metric system increasingly dominates our world, there are plenty of people who still think of all sorts of measurements in imperial terms. They are not so many metres and centimetres high, but measure themselves in feet and inches. Similarly, kilograms mean nothing to them, but stones and pounds leave them fretting every time they stand on the bathroom scales.

Quiz 59

In this quiz, the aim is to convert common metric measures of distance and mass into their imperial equivalents and vice versa. See how you get on.

Metric	***Imperial***
1 kilogram	____ pounds
1 tonne	____ tons
1 centimetre	____ inches
1 metre	____ feet and ____ inches
1 kilometre	____ miles
1 square metre	____ square yards
1 hectare	____ acres
1 litre	____ pints

Quiz 60

Now try it the other way round.

Imperial	***Metric***
1 ounce	____ grams
1 pound	____ kilograms
1 stone	____ kilograms
1 ton	____ kilograms
1 inch	____ centimetres
1 foot	____ metres
1 yard	____ centimetres
1 mile	____ kilometres
1 acre	____ hectares
1 square foot	____ square metres
1 pint	____ millilitres

Answers on pages 193–4

Off the Scale

In the metric world, we tend to measure temperatures using the Celsius (also known as Centigrade) scale. However, plenty of people still use the Fahrenheit scale (Incidentally, Daniel Fahrenheit used the temperature of his wife's armpit as a key value in devising his scale!).

Unfortunately, the formulae needed to translate from one temperature scale to another are quite complex but they do make for good maths:

- *To convert from Celsius to Fahrenheit => Temperature °F = (Temperature °C x $^9/_5$) + 32*
- *To convert from Fahrenheit to Celsius => Temperature °C = (Temperature °F – 32) x $^5/_9$*
- *To convert from Celsius to kelvin, simply add 273° (or subtract 273° to go from kelvin to Celsius).*
- *Going to or from Fahrenheit to kelvin is way too complex so all right-thinking people do a double conversion using Celsius.*

Quiz 61

1) Convert 1°C into Fahrenheit.
2) Convert 1°F into Celsius.
3) Convert 20°C into Fahrenheit.
4) Convert 59°F into Celsius.
5) Convert -5°C into Fahrenheit.
6) Convert -13°F into Celsius.
7) Convert 6°C into kelvin.
8) Convert 138°K into Celsius.
9) Convert 77°F into kelvin.
10) Convert 278°K into Fahrenheit.

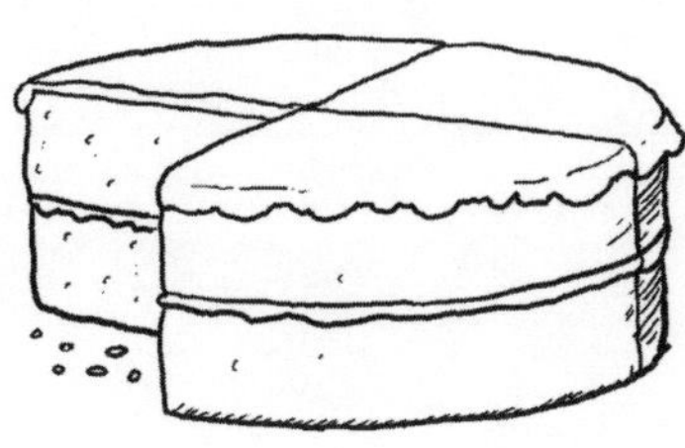

Answers on page 194

Mine's Bigger than Yours

In the following quiz, you need to work out whether the imperial or metric measure is larger.

Quiz 62

1) Tommy had a bag of apples weighing 4.5lb, while Geoffrey's weighed 2kg. Whose bag of apples was bigger?
2) The red sports car has a top speed of 250km/hour and the blue sports car can achieve 150 miles/hour. Which car is faster?
3) In a boxing match, the fighter in the red corner weighed 92kg and the boxer in the blue corner weighed 14 stone and 10 pounds. Which corner had the heaviest fighter?
4) Farmer Silage had a farm of 37.75 acres while Farmer Straw's covered 15 hectares. Who had the bigger farm?
5) Lawrence carpeted the lime bedroom with 26 square yards of carpet and used a further 22 square metres in the puce bedroom. Which room was bigger?
6) It was 27°C in New York and 79°F in Boston. Which city was warmer?

Answers on page 195

Quiz 63: Data and Measurements Crammer

1) 10 people are surveyed about the number of exam passes they have achieved. Their results are as follows:
   0, 2, 4, 6, 7, 8, 8, 9, 10, 12
   Give the average number of passes held by the respondents expressed as the mean, median and mode.
2) A doctor carries out a study of 45 patients and chooses to express his data in a pie chart. How many degrees of the pie would each patient take up?
3) What sort of correlation would you expect to see in a scattergraph plotting time spent playing computer games in an evening against time spent doing homework?
4) How many watts are there in a gigawatt?
5) Which weighs more: 10kg of potatoes or 20lb of carrots?
6) Tony is 6 foot 2 inches tall and Joe is 162cm. Who is taller?
7) Janet drank 3½ litres of beer while Geraldine drank 6½ pints. Who is likely to be drunker?
8) The temperature in Paris was 18°C. What is that in Fahrenheit?

Answers on page 195

LETTERS BEGIN: ALGEBRA

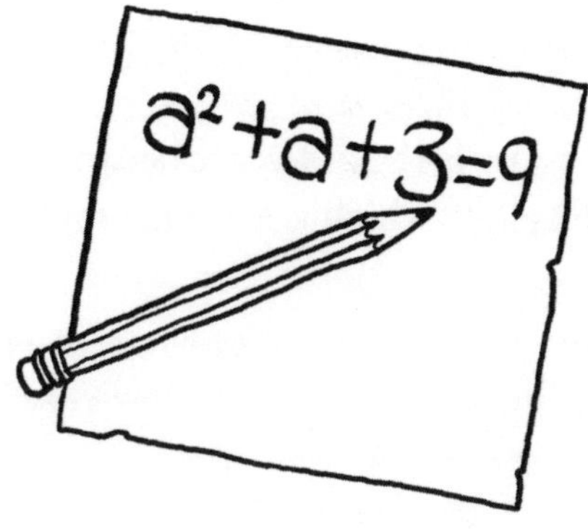

Algebra is guaranteed to send a shiver down the spine of anyone who's sat at the back of the classroom with the maths teacher's words flying straight over their head. But that need not be the case.

In algebra, we use letters in place of certain unknown numbers. We do this when we have an equation or a formula where one or more elements have an unknown value but a fixed relationship with the other values. By the way, you can use pretty much whichever letter you want to replace a number.

Secret Formula

It is worth being aware of the difference between a straightforward equation and a formula (which is a special sort of equation).

- *In an equation, any letter has only a limited number of possible values (indeed, often only one potential value). For instance, in the equation* $5 + a = 9$*, a can only equal 4.*
- *In a formula, though, there can be a whole range of values.*
    - *For instance, the formula for calculating speed (s) involves dividing distance travelled (d) by time taken (t). Or* $s = d \div t$*. This equation means that as long as we know the distance and the time (whatever values those might be), we have the tools to work out speed. For instance:*
        - $d = 10$*km* $t = 0.5$ *hours*
        - $s = d \div t = 10 \div 0.5 = 20$*km/hour*
    - *Equally, d might be 100km or 5,000km or 35 metres, and t might be 3 hours or 6 hours or 6 weeks, and we can still calculate s.*

Quiz 64

Which of the following are formulae and which are equations?

1) $3 + a = 143$
2) $e = mc^2$
3) $q - 12 = -14$
4) $z + 7 - 9 = 21$
5) $c = d \times 3.14$
6) $c = 12.5 \times 3.14$

Answers on page 196

The Language of Letters

The world of algebra has a language of its own. Here are some of the key terms:

- *A variable is a symbol given to an unknown value. It is usually represented by a letter, e.g. a, x or y.*

- *A coefficient is a number placed in front of a variable, e.g. 3 is the coefficient in 3a.*

- *A term refers to a constant, e.g. 3, 10, 1/4, as well as the product of a coefficient and a variable, e.g. 3a, and the product of two or more variables, e.g. xy. In algebraic terms like this, there is no need for a multiplication sign.*

- *Like terms are two or more terms that have the same unknown (or unknowns) with the same power. For example,* a^2 *and* $3a^2$ *are like terms, as are a and 3a (in this latter example the unknown, a, has the same power, 1, in each case).*

- *Unlike terms are any terms that do not have the same unknown (or unknowns) with the same power. For example, a and* a^2 *are unlike terms, as are a and b.*

- *An expression is made up of one or more terms, e.g.* $3a + xy$*.*

- *An equation is made up of two expressions separated by an equal sign. The expressions have the same value, e.g.* $3a + xy = 12a + 4xy$*.*

Quiz 65

1) In the following equation, identify any and all expressions: $3b = 21$
2) In the following equation, identify any and all terms: $14 - 6w = 5w - 8$
3) In the following equation, identify any and all coefficients and variables: $14 - 6w = 5w - 8$
4) In the following equation, identify any and all like terms: $t + t^2 - 4 = 2t + 2$
5) In the following equation, identify any and all pairs of unlike terms: $2p + p^2 = 3p + 20$

Answers on page 196

Starting Simple

It is fair to say that some algebra is enormously complicated but by and large we can leave that sort of thing to the boffins. But getting to grips with the basics is within the grasp of virtually all of us.

There are a couple of basic rules always to keep in mind:

- *Ultimately we are looking to isolate a particular letter on one side of the = sign so as to work out its value using the sum (hopefully a simple one) on the other side of the = sign.*

- *To do this, we have to simplify our expressions. This is guided by one key tenet – what you do to one side of an equation you must also do to the other side to keep it balanced.*

Let's take a really easy example to illustrate this.

- *Tony and Angus have 9 toy cars between them. We know that 5 belong to Tony but we are unsure how many belong to Angus (yes, I know we could hazard a pretty good guess but stay with me on this one!).*

- *We decide to write an algebraic expression, replacing the unknown quantity (the number of cars belonging to Angus) with a letter (let's choose a).*
- *We can write the following sum:* $a + 5 = 9$.
- *In order to isolate the a, we need to get rid of that 5. To do so, we simply minus 5. Having done that in the expression on the left, we must also do it on the right. Our equation is now* $a + 5 - 5 = 9 - 5$.
- *We can further simplify this to* $a = 9 - 5$, *which gives us our answer: Angus has 4 cars.*

Quiz 66

OK, now to test yourself with some reasonably straightforward equations. In each case, you need to identify the unknown factor a–i. Before you do so, bear in mind the BIDMAS rules on page 42.

1) $a + 7 = 25$
2) $28 - c = 12$
3) $8 + d + 2 = 19$
4) $48 - 11 - e = 32$
5) $19 + f - 7 = 28$
6) $24 + 36 - 5 - g = 1$
7) $112 - h + 43 = 7 \times 14$
8) $45 \div 9 + i = 72 \div 3$

Answers on page 197

Multiply, Divide and Conquer

In this quiz, you will need to start multiplying and dividing both expressions to isolate the variables. It is not as scary as it sounds!

Quiz 67

Work out the unknown factor in each case.

1) $3h = 24$
2) $7i - 14 = 49$
3) $j/3 = 33$
4) $k/12 + 38 = 50$
5) $3e + 11 = -20/2$
6) $3f \div 9 = -2$

Balancing the Books

The more complicated equations get, the harder we have to work to simplify them. Here are a few methods you can use to this end.

- *Simplify like terms. Any two terms, whether like or unlike, combine to form a single term when multiplied or divided. But when added or subtracted, only like terms will simplify to a singly term (by adding or subtracting their coefficients appropriately).*

- *Simplify indices. If you are multiplying unknowns or numbers that are the same, you can combine their indices by adding them together. For instance, $2^2 \times 2^3$ is the same as 2^{2+3} or 2^5.*

- *If there are coefficients attached to the unknowns, add the indices*

Answers on page 198

as above and multiply the coefficients. Thus, $5a^2 \times 3a^3$ *becomes* $(5 \times 3) \times a^{2+3}$*, or* $15a^5$.

- *Dividing with indices works in completely the opposite way. That is to say, indices are subtracted and coefficients divided. So, for example:*
  - $a^5 \div a^3$ *is the same as* a^{5-3}*, or* a^2*. (And note, you can have negative indices!)*
  - $16a^5 \div 4a^3$ *is the same as* $(16 \div 4)\ a^{5-3}$*, or* $4a^2$*.*

Quiz 68

In this quiz, use the above methods to simplify the equations (but don't worry about solving them):

1) $5a - 2a + 1 = 7$
2) $4b + 2b^2 + 3b = 22$
3) $c^2 \times 3^2 \times c^2 \times 3^2 = 7{,}776$
4) $2d^2 \times 2d^3 = 972$
5) $e^5 \div e = 81$
6) $6f^5 \div 3f^3 = 18$

Answers on page 198

The Bracket Racket

To add to the algebra fun, you will sometimes find that your unknown is nestled within some brackets. In such cases, you will need to liberate the unknown from its shackles in a process known as expanding brackets.

To do this, you need to multiply everything inside the brackets by whatever sits immediately outside them. For instance, say your equation is 5(a + 2) = 30:

- *Take 5(a + 2) and turn it into (5 x a) + (5 x 2).*
- *Don't worry – you haven't just added another set of brackets to the torment. Instead, after a quick calculation you end up with 5a + 10.*
- *So our equation now reads 5a + 10 = 30. Therefore, 5a = 30 – 10 = 20.*
- *If 5a = 20, a = $^{20}/_{5}$ = 4. Equation solved and brackets defeated!*
- *Do remember to keep any operation signs (e.g. +, –) with the figure to its right. Hence, 5(a – 2) becomes (5 x a) + (5 x -2) and can be further simplified to 5a – 10.*

Quiz 69

1) 2(a + 4) = 14
2) 6(b – 6) = 42
3) 4(c + 3) – 8 = 32
4) –3(d + 4) = –39
5) 3(e + 2) = –15
6) –6(f – 3) = 42

Answers on page 199

Into the Unknowns

Some equations have unknowns on both sides of the = sign. In such cases the aim is to balance the equation so there are no unknowns on one side. Just remember, what you do to one side, you must do to the other (and if there are brackets involved, you will need to expand them).

- *Let's look at the sum $4 + 2a = 3a + 1$. To get rid of the unknown on the left, we subtract $2a$, so must also do the same to the expression on the right. The equation is now $4 = a + 1$. This, you will quickly see, means that $a = 4 - 1 = 3$.*

- *In this example, $3b + 16 = 4(b + 3)$. If we expand the brackets, we end up with $3b + 16 = 4b + 12$. A further simplification reveals that $16 = b + 12$, so $b = 16 - 12 = 4$.*

Quiz 70

Work out the value of the unknown in each of the following equations.

1) $6a + 2 = 4a + 10$
2) $12b - 5 = 10b + 9$
3) $3c + 22 = 6c - 5$
4) $2(d + 2) = 3d - 1$
5) $e + 49 = 3(e + 1)$
6) $3f + 2 = 4f + 4$
7) $18 - g = -3g + 4$
8) $4h - 5 = 3(h + 3)$

Answers on page 200

Word Problems

In this exercise, derive algebraic equations from the word questions to solve the problem.

Quiz 71

1) Add 10 years to my age and you get 45. How old am I?
2) 3 times my front door number is 126. What is my door number?
3) If you divide my father's age by 3 and add 14, you get 40. How old is my father?
4) It you take the average test score of the pupils in my maths class and add 5 to it, and then divide that figure by 3, you get 28. What was the average test score?
5) If you take the number of students in the class, multiply that figure by 3 and add 17, you get the same result as when you multiply the class size by 4 and take away 2. How many students are there in the class?

Answers on page 201

X, Y & Z Factor

Factorizing is the process of making brackets, something done to make formulae easier to deal with or to help solve equations where there are unknowns with powers.

- *As we already know, a factor is a number that exactly multiplies into another number. So if we were to factorize 14 + 21 = 35, we would end up with 7(2 + 3) = 35. Of course, you would never do this in practice with simple numbers but it illustrates the process.*

- *Let's now look at an algebraic equation.* $8a + 10 = 66$. *We know that 2 is a factor of both 8 and 10 so we can rewrite the equation as* $2(4a + 5) = 66$.

- *We can also factorize expressions that include unknowns with a power. For instance,* $8a^2$ *may be expressed as* $2a(4a)$ *or, indeed,* $4a(2a)$.

- *Taking this a stage further, if our equation was, say,* $8a^2 + 2a = 78$, *we could rewrite it as* $2a(4a + 1) = 78$.

- *You can even factorize expressions where there is more than 1 unknown. For instance,* $4ab + 4a^2b^3$ *could be written as* $2ab(2 + 2ab^2)$.

Quiz 72

Look at the following equations and have a go at factorizing where you can (but don't worry about solving them).

1) $64 + 24 = 88$
2) $9a + 15 = 51$
3) $24b - 36 = 12$
4) $c^2 + 5c = 66$
5) $4d + 21 = 41$
6) $3f^2 + 6f = 72$

Answers on page 201

Quad-Squad

So that's the theory of factorizing in algebra but what actually is the *point*? Well, so far all the equations we have looked at have been linear. That is to say, they have contained unknowns that have been multiplied, divided, added to or subtracted from by a number.

However, there are also things known as quadratic equations, in which an unknown is squared. These can be pretty tough equations to crack but factorizing goes a long way to making it possible. More of that later. First a few basics.

Quadratic equations have a standard format like this: $ax^2 + bx + c = 0$. The a, b and c are coefficients while the x is the unknown. b and c might have the value 0, and quadratics can also take on several disguises (too complex to explain here, alas) but essentially a quadratic equation can always be brought round to this format.

Quiz 73

Decide which of the following are quadratic equations (e.g. can be made to fit the format above) and which are not.

1) $2x^2 + 3x - 6 = 0$
2) $6x + 3x - 2 = 0$
3) $2x^2 + 3x = -6$
4) $7x^2 + 5 - 9 = 0$
5) $7x^2 + 5x - 4x = 8$

Answers on page 202

Quiz 74

In this quiz, rewrite the following equations so they fit the classic quadratic format.

1) $3x^2 + 6x = 7$
2) $x^2 - 2 = -3x$
3) $x^2 = 4x - 2$
4) $3(x^2 - 3x) = 8$
5) $x(x + 4) = -6$

Answers on page 202

Double the Trouble

The tricky thing about quadratic equations is that by their very nature they have two answers. How can that be, you ask? Let's consider an apparently very simple equation: $a^2 = 4$. Using our impeccable knowledge of square numbers, the answer is clearly 2, isn't it? But wait a minute. We know that if you multiply a negative by a negative you get a positive. So $(-2)^2$ also equals 4. Therefore, a might be either 2 or -2. There is a special symbol to indicate that the answer may be either positive or negative. The answer to $a^2 = 4$ is that $a = \pm 2$.

So where does factorizing come into play. Consider this quadratic equation: $d^2 + d = 20$. If we carry out some normal simplification, we might end up with $d^2 + d - d = 20 - d$. Now we have unknowns on all sides of the = sign and find ourselves in a pretty mess. So instead we factorize it like this: $d(d + 1) = 20$. We make an educated guess that $d = 4$, and put it into the equation: $4^2 + 4 = 16 + 4 = 20$.

Except, we know that quadratic equations have two possible answers. We could try -4 as the alternative answer but the sums don't add up. How about -5 then? $-5^2 + -5 = 25 - 5 = 20$. Therefore, d = 4 or -5. Success. But for real mathematicians, our approach was a bit hit and miss. You can, however, reach your answers more methodically.

- *Start by getting your quadratic equation into the classic format. Thus, for instance, $e^2 + 4e + 2 = 7$ becomes $e^2 + 4e - 5 = 0$ (note, we have taken 7 off both sides to get a 0 on one side of the brackets.*

- *Now we want to factorize the expression on the left. To do this we end up with a pair of brackets, each containing c ± a number. In this case, we get $(e - 1)(e + 5)$.*

- *How do we end up with −1 and +5 in the brackets? There is a simple general rule to follow here. Remember our skeleton equation $ax^2 + bx + c$? We are looking for the two numbers that add up to give b (here the 4 of 4e) and multiply to equal c (here −5).*

- *A bit of arithmetic tells us that the only whole numbers that multiply to give −5 are 1 and −5, or 5 and −1. And it is only the latter pair that when added together gives 4.*

Quiz 75

Let's briefly interrupt solving our specimen quadratic equation. In each case below, come up with the pair of figures that add together to give the first number and multiply to give the second.

1) 10, 21
2) 15, 54
3) –1, –6
4) 4, –32
5) 49, 220
6) 47, 532

Right, back to solving our equation. To recap:

- *$e^2 + 4e + 2 = 7$ becomes*
- *$e^2 + 4e - 5 = 0$*
- *Our factors of b and c are –1 and 5, hence we get to*
- *$(e - 1)(e + 5) = 0$*
- *It follows that either $e - 1 = 0$ or $e + 5 = 0$.*
- *e is therefore worth +1 or –5.*

Answers on page 202

Quiz 76

Got all that? Now it's your turn. See how you get on solving the following simple (it's all relative!) quadratic equations.

1) $a^2 - 10a + 25 = 0$
2) $b^2 + 20b + 91 = 0$
3) $c^2 - 2c - 8 = 0$
4) $d^2 + 3d - 54 = 0$
5) $e^2 + 21e = -68$

Answers on page 203

All at Once

Simultaneous equations are two equations that are both true at the same time and which contain two or more unknowns to be found.

The simplest way to solve them is to add or subtract one from the other to eliminate one of the unknowns from the equation.

- *Take the following equations:*
  - $2a + b = 17$
  - $3a - b = 18$
    - *If we add the equations together we get $5a = 35$ (since the $+b$ and $-b$ cancel each other out). A quick sum reveals that a must thus equal $35 \div 5$, which equals 7.*
    - *Plug 7 into either equation in place of a and we discover that b must then equal 3.*

But what if a simple addition or subtraction doesn't eliminate either of the unknowns?

- *Simply factor up one of the equations until you can. Look at the following example:*
  - $4c + 3d = 47$
  - $2c - d = 1$
- *If we add them together we get $6c + 2d = 48$. If we subtract one from the other we get $2c + 4d = 46$. Neither helps us.*
- *So, using the lowest common denominator, we can turn the second equation into $4c - 2d = 2$.*
- *Now if we subtract we end up with $5d = 45$. Therefore, $d = 45 \div 5 = 9$. By plugging this 9 into either equation we can then calculate that c must equal 5.*

Quiz 77

Have a go at answering the following simultaneous equations.

1) If $3a + b = 22$ and $4a + 6b = 62$, what are the values of a and b?
2) If $2c - d = 12$ and $4c + d = 42$, what are the values of c and d?
3) If you add Jack's age to Beryl's age, you get 84. If you take Beryl's age from Jack's you get 12. How old are Jack and Beryl?
4) If you take Pete's hourly wage and add it to twice Sylvia's hourly wage, you get £20.50. If you subtract Sylvia's hourly wage doubled from Pete's hourly wage tripled, you get £5.50. How much do Pete and Sylvia earn per hour?
5) Will buys 3 sandwiches and 2 chocolate bars for £5.70. Helena buys 2 sandwiches and 6 chocolate bars for £7.30. What is the price of a sandwich and how much does a chocolate bar cost?

Answers on page 203

Quiz 78: Algebra Crammer

1) If $27 + 6 - a = 15 - 6$, what is a worth?
2) If $4b - 12 = 60$, what is b worth?
3) If $c/6 + 18 = 25$, what is c worth?
4) Solve $4(2d - 3) = 124$.
5) $12 + 2e = 3e - 17$, what is e worth?
6) If $3f + 48 = -5f$, what is f worth?
7) Solve the quadratic equation, $g^2 + 8g = -15$.
8) Jill buys 3 cups of tea plus a hot chocolate for £3.90. Jack buys 2 hot chocolates and a cup of tea for £3.55. How much is a cup of tea and how much for a hot chocolate?

Answers on page 204

TAKING SHAPE: GEOMETRY AND TRIGONOMETRY

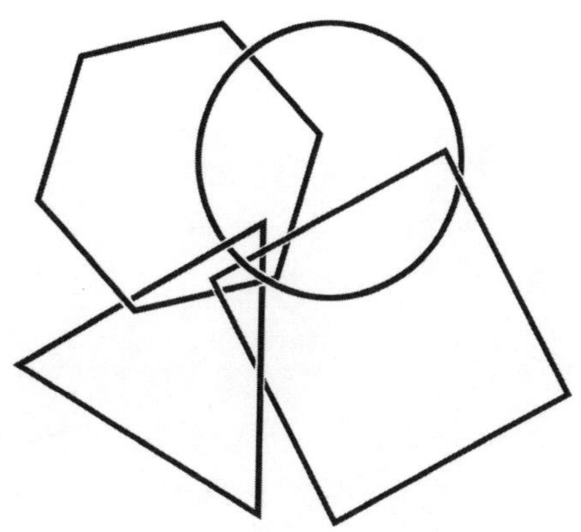

From the Ancient Greek for 'Earth measuring', geometry is the branch of maths that looks at the interaction between shapes, lines and angles.

Coming at It from a Different Angle

Angles occur when two lines touch each other or cross (and the point where they meet or cross is called a vertex). There are 360° in a complete revolution and there are four major classifications of angles:

- *Right angles*
- *Acute angles*
- *Obtuse angles*
- *Reflex angles*

Quiz 79

Can you match each of the angles below to its correct term and definition?

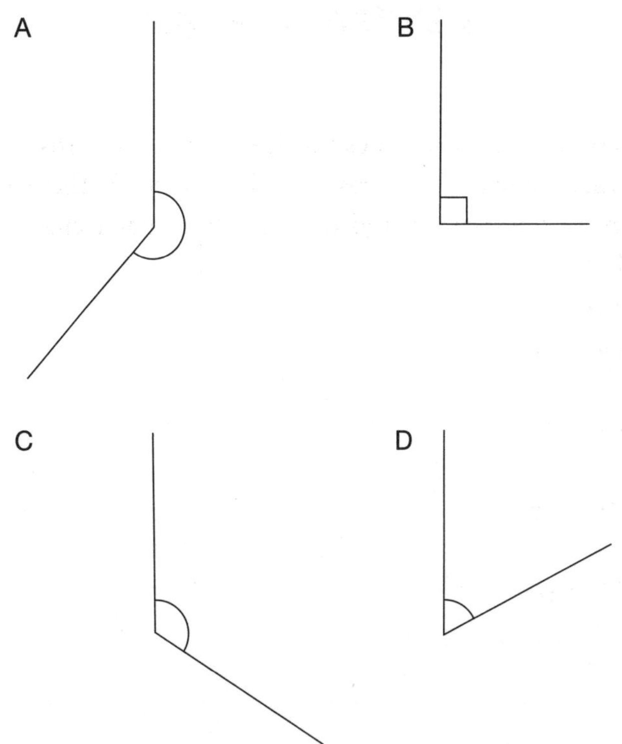

Term	*Definition*	*Image*
1) Right angle	i) An angle less than 90°	A)
2) Acute angle	ii) An angle greater than 180°	B)
3) Obtuse angle	iii) An angle of 90°	C)
4) Reflex angle	iv) An angle between 90° and 180°	D)

Answers on page 204

Dead Parrots?

There is an old joke about polygons and dead parrots but polygons are, in fact, two-dimensional shapes consisting of multiple straight lines that connect up to produce a closed shape.

Quiz 80

In the table below, the column on the left describes the number of sides in different polygons. In the column on the right, are you able to name the shapes?

Number of Sides	Name of Polygon
3	
4	
5	
6	
7	
8	
9	
10	
11	
12	
20	
50	
100	

Answers on page 205

All Square (and Triangular)

Triangles and quadrilaterals come in various types but the internal angles of a triangle will always total 180° while those of a quadrilateral always add up to 360°.

Quiz 81

In this quiz, can you correctly label the following triangles and quadrilaterals? (Dashes through straight lines indicate sides of equal length. Angles marked with the same letter indicate that they are equal.)

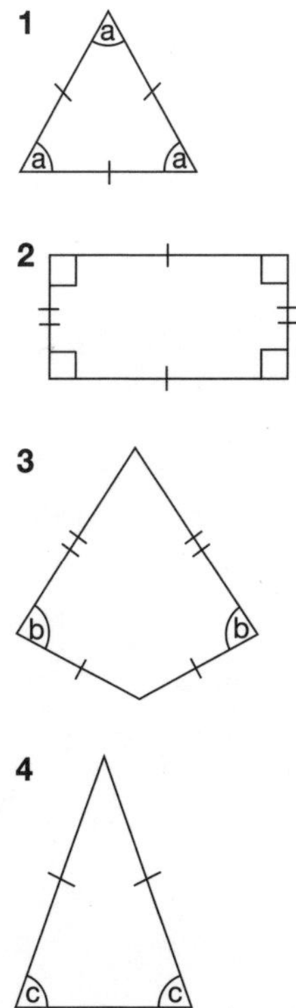

Answers on page 205

5

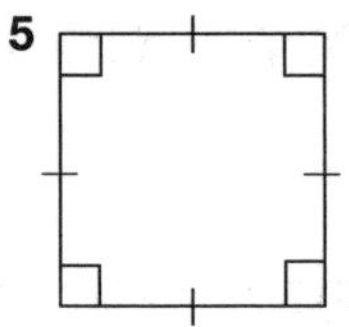

6

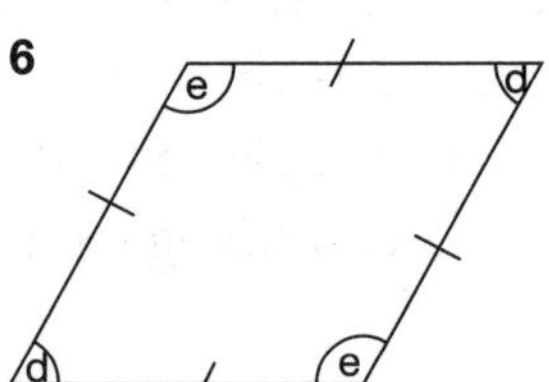

7

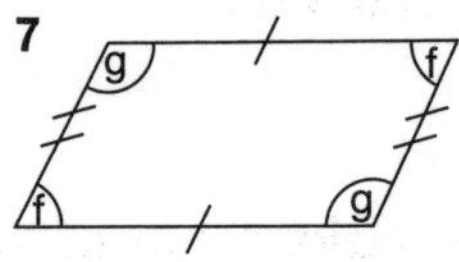

8

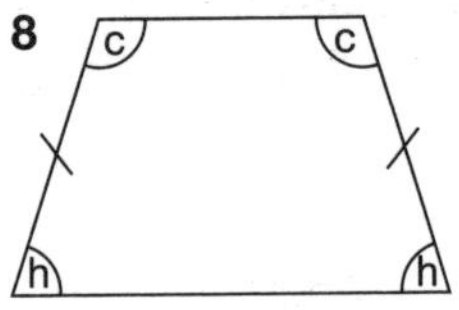

9

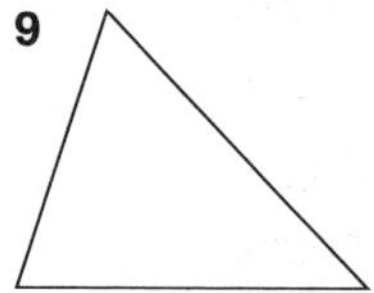

Labels:

a) Isosceles triangle
b) Rhombus
c) Equilateral triangle
d) Scalene triangle
e) Parallelogram
f) Rectangle
g) Kite
h) Square
i) Trapezium

Answers on page 205

Inside Information

All the interior angles (that is to say, the angles within a shape) of perfectly regular polygons have equal values. For instance, in an equilateral triangle, each angle has the same value and since we know the internal angles of a triangle total 180°, we know that each of the angles in an equilateral triangle must be worth 60°.

Quiz 82

Below are listed the names of various polygons. Assuming they are perfectly regular examples, can you give the value of each internal angle in each case? (And a massive pat on the back if you get them all!)

1) Quadrilateral	Angle: ______________°
2) Pentagon	Angle: ______________°
3) Hexagon	Angle: ______________°
4) Heptagon	Angle: ______________°
5) Octagon	Angle: ______________°
6) Nonagon	Angle: ______________°
7) Decagon	Angle: ______________°
8) Hendecagon	Angle: ______________°
9) Dodecagon	Angle: ______________°
10) Icosagon	Angle: ______________°

Answers on page 206

A Greater or Lesser Degree

There are a few facts about angles that can help us to work out the values of unknown angles.

- *Angles on a straight line add up to 180°.*

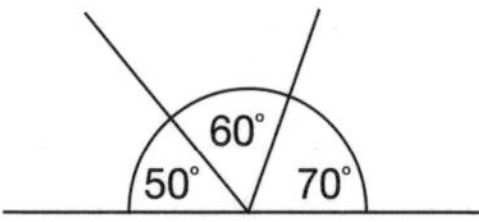

- *Angles opposite each other across a vertex (known as vertically opposite angles) are always equal.*

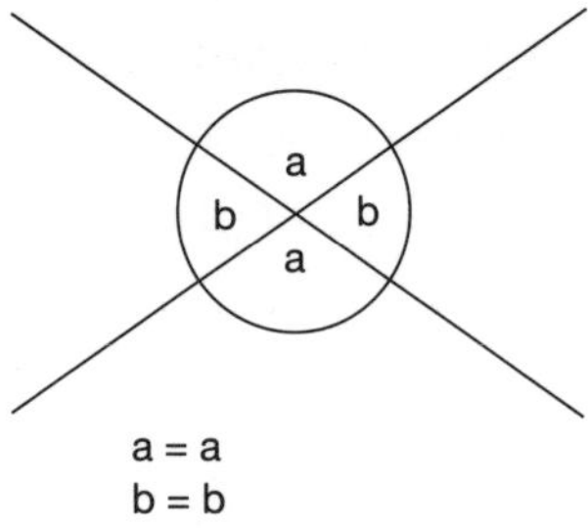

a = a
b = b

- *When a pair of parallel lines is intersected by a third line, we end up with 2 sets of identical angles.*

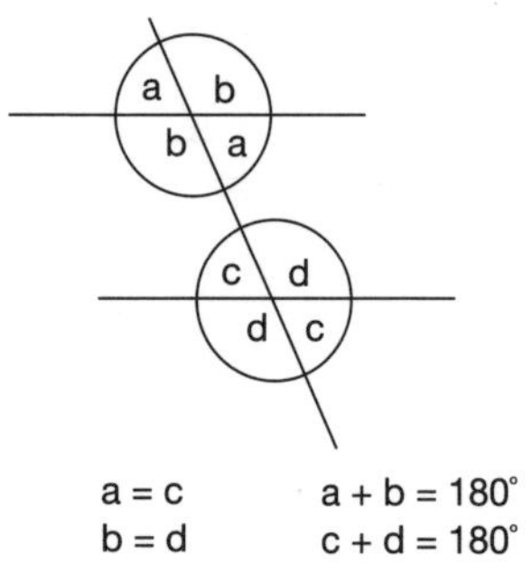

a = c　　a + b = 180°
b = d　　c + d = 180°

  - *These include 2 pairs of vertically opposite angles. The angles marked a are equal to c. Similarly, the angles marked b are equal to those marked d.*
  - *We also have angles on a straight line, so we know that angles a + b = 180°, as do angles c + d.*

- *It doesn't end there. We also have alternate angles, which are those angles that are in different sets but which if they were in the same set would be vertically opposite. They are, thus, equal. In the example below, for instance, angles e and f are alternate.*

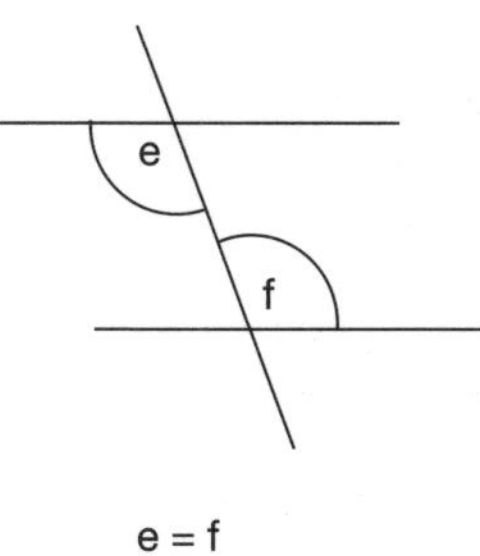

e = f

- *We also have co-interior angles, which are those from different sets that if they were in the same set would lie on the same straight line, and thus add up to 180°. In the example below, for instance, angles g and h are co-interiors.*

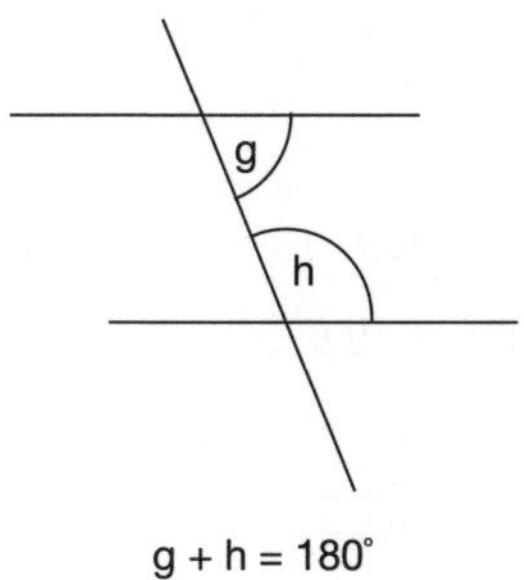

g + h = 180°

Quiz 83

Look at the following diagrams and calculate the values of the missing angles.

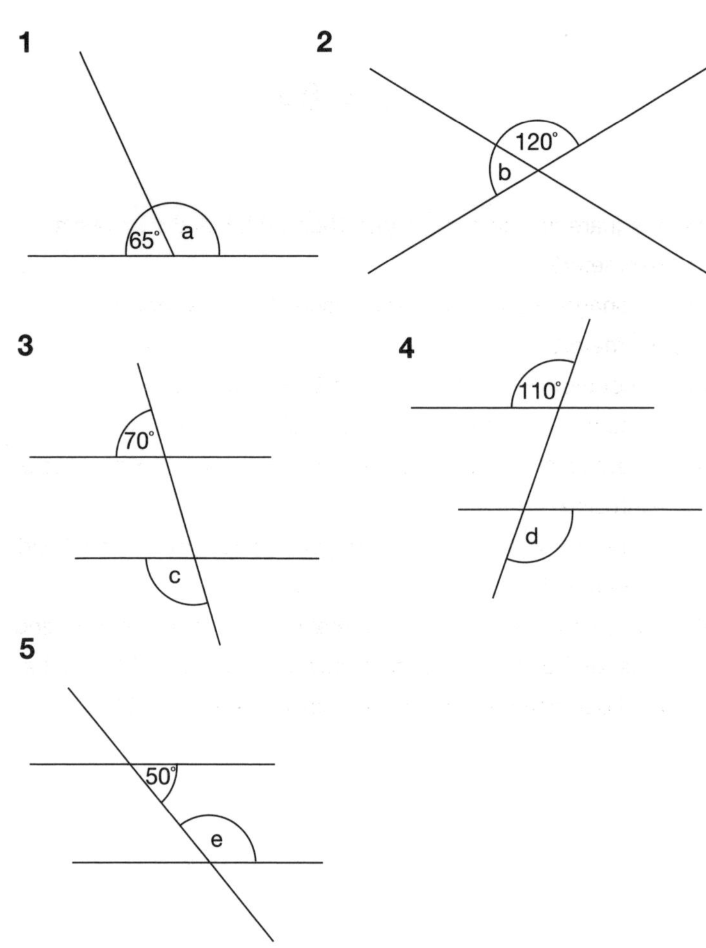

Answers on page 206

Going Round the Houses

The perimeter of a shape is the total length around it. The perimeter of a square, for instance, is the combined total of the length of its four sides.

Quiz 84

1) A square has sides of length 10cm. What is the square's perimeter?
2) A nonagon has sides that measure 13cm. What is its perimeter?
3) A rectangle has a short side of 24cm and a long side 1½ times as big. What is the rectangle's perimeter?
4) An equilateral triangle has a perimeter of 51cm. How long is each side?
5) If a regular polygon has a perimeter of 192cm, and each side is 16cm in length, what is the shape?
6) Three squares and an equilateral triangle are laid end to end. Each of the shapes has sides measuring 16cm. What is the total perimeter around the shapes laid end to end?

Answers on page 207

Access All Areas

Area is the amount of flat space something takes up. It is, if you like, an object's 'footprint'. Here are a few useful equations for calculating area.

- *Working out the area of a square or a rectangle is as straightforward as it gets. Area (A) = Width (W) x Height (H). So if we have a rectangle with a width of 10cm and a height of 7cm, its area is 10 x 7 = 70cm^2.*

- *The area of a triangle is exactly half the area of the imaginary rectangle into which it fits. This handy facts allows us to use the following equation: Area = ½ x Base x Height. Remember, though, that the height of a triangle is not the same as its longest side. The height is measured from the middle of the base to the tip of the triangle.*

- *With a parallelogram, Area = Base x Height (though, again, remember that the height is not the same as the length of one side.*

- *For trapeziums, the equation is more complicated:*

  $$\text{Area} = \frac{(a + b) \times \text{Height}}{2}$$

  In this equation, a and b are the two parallel lines in a trapezium.

- *For a kite, imagine drawing two lines connecting opposing vertices. These are the diagonals of a kite. Thus Area = $^1/_2$ x diagonal a x diagonal b.*

- *For all other polygons, remember that they can always be divided into triangles, rectangles or a combination of the two. It's then a case of working out the area of each sub-shape and adding the totals together.*

Quiz 85

Work out the area of each of the following shapes.

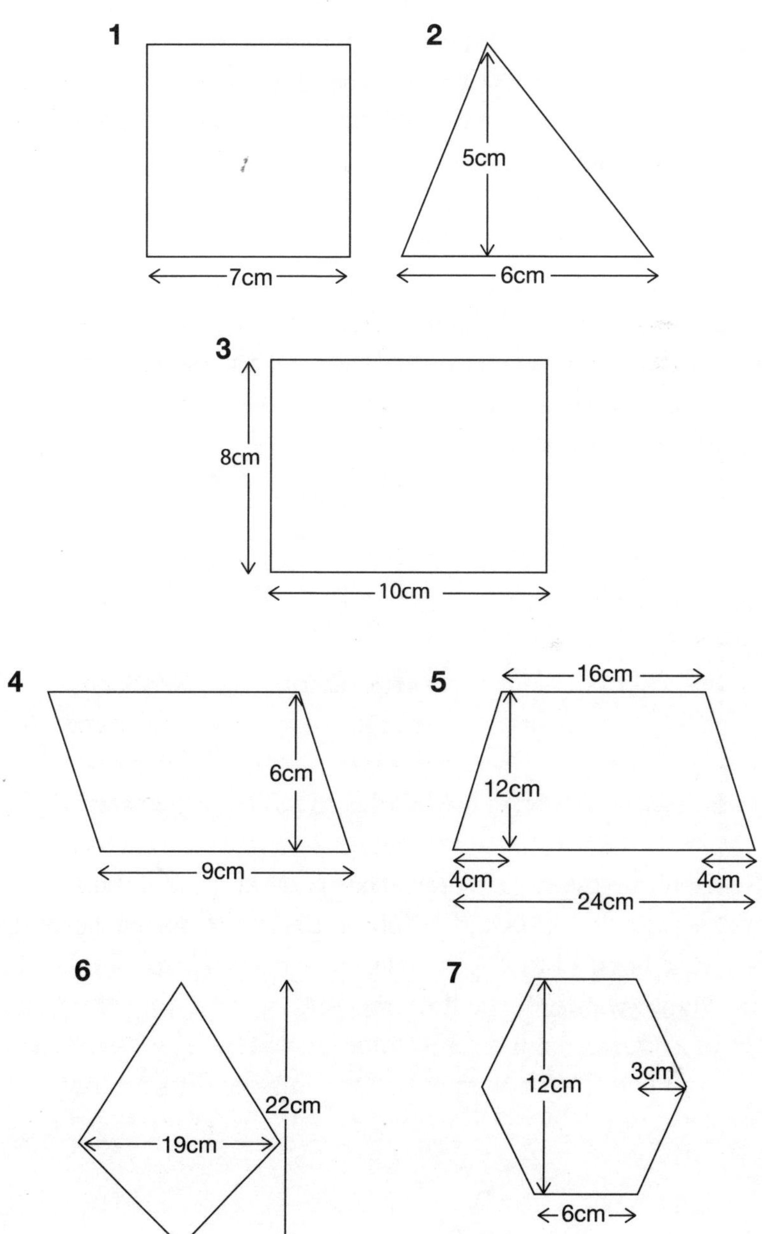

Answers on page 207

Don't Let the Hypotenuse Confuse

Pythagoras, who possessed one of the greatest minds of Ancient Greece, is now best known for the discovery he made about the qualities of right-angled triangles (which is to say, a triangle with one corner made up of two sides meeting at 90° to each other).

This theorem, which has delighted and confused generations of school children in equal measure, states: The square of the hypotenuse of a right-angled triangle is equal to the sum of the squares of the other two sides.

The hypotenuse is the name given to the longest side of a right-angled triangle, which you'll always be able to find directly opposite the right angle itself. If we call the hypotenuse h and the other two sides a and b, we can represent the theorem with the following equation:

$$h^2 = a^2 + b^2$$

Quiz 86

Armed with all that, can you solve the following posers?

Geoff is the head gardener at a great mansion, Hypotenuse House, and he is trying to follow the plans drawn up by his eccentric boss, Lord Rightangle. True to his name, the Lord of the Manor wants all the flowerbeds to be right-angled triangles but in each case he has forgotten to include vital measurements. Can you help Geoff work out all the missing dimensions?

Answers on page 208

1) How long is side h?

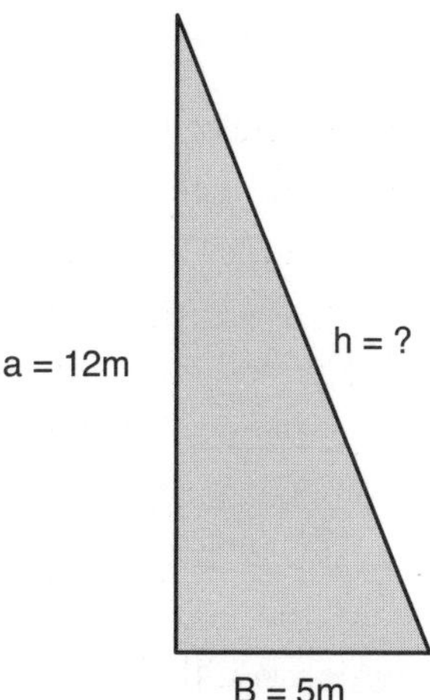

2) How long is side b?

a = 4m
h = 5m
B = ?

3) How long is side a?

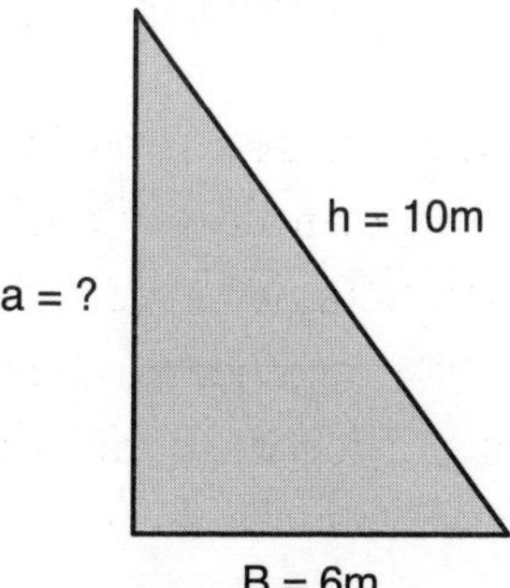

Answers on page 208

4) Lord Rightangle wants a square summerhouse built along the long edge of this triangle. What will its area be?

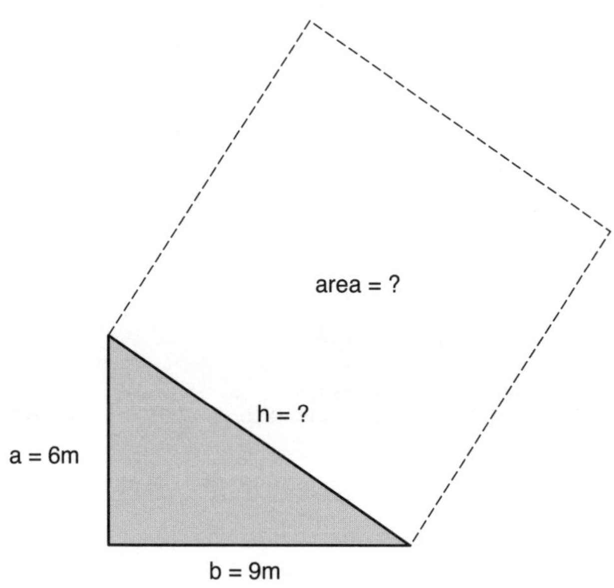

Answers on page 208

Around and About

Circles are technically polygons, but with so many short sides linked together that it looks like one long, continuous curved side without beginning or end. Being a bit different, circles have some distinct vocabulary of their own.

Quiz 87

Match the following terms relating to circles with their correct definition.

1)	Circumference	a)	The distance from the centre of a circle to its edge
2)	Radius	b)	A value (beginning 3.14159) key to calculating circumference
3)	Diameter	c)	The perimeter of a circle
4)	Pi	d)	The distance across the whole of a circle via its centre point

Answers on page 208

The Life of Pi

Pi, which runs to an infinite number of decimal places, is one of the most magical numbers in mathematics as it unlocks so many of the mysteries of circles. For our purposes, let's leave its numerical value at 3.14159 and use the symbol π to denote it. Here are some more equations to calculate various measurements related to circles (a = area; c = circumference; d = diameter; r = radius):

- $\pi = c \div d$
- $c = \pi \times d$
- $c = 2\pi r$
- $a = \pi r^2$

Quiz 88

With those in your back pocket, answer the following questions:

1) What is the radius of a circle of diameter 23.5cm?
2) Calculate the diameter of a circle of radius 59.27cm.
3) Give the circumference of a circle of radius 14cm.
4) Calculate the area of a circle of radius 35cm.
5) Give the area of a circle of diameter 94.5cm.
6) If a circle has a circumference of 75.4cm, what is its radius?

Answers on page 209

Solids

Solids are 3-dimensional shapes. The main categories of solids are:

- *Polyhedra – solids with flat faces:*
  - *Prisms – a stretched shape which will have the same cross-section across its length*
  - *Pyramids – made by connecting a base to an apex (that is to say, a point)*
  - *Platonic solid – where each face is the same regular polygon, with the same number of polygons meeting at each vertex.*
- *Non-polyhedra – solids with any surface that is not flat:*
  - *Sphere – a round, ball shape*
  - *Cylinder – a curved prism*
  - *Cone – essentially a curved pyramid*
  - *Torus – a doughnut-shaped solid*

Quiz 89

Put the right labels with the corresponding solids below:

Labels

a) Octahedron
b) Tetrahedron
c) Sphere
d) Pentagonal prism
e) Cylinder
f) Square pyramid
g) Cube
h) Cuboid
i) Cone
j) Torus

1

2

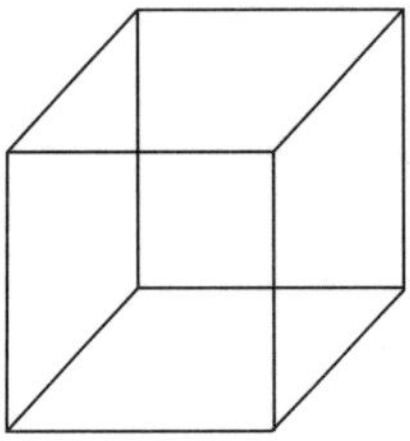

3

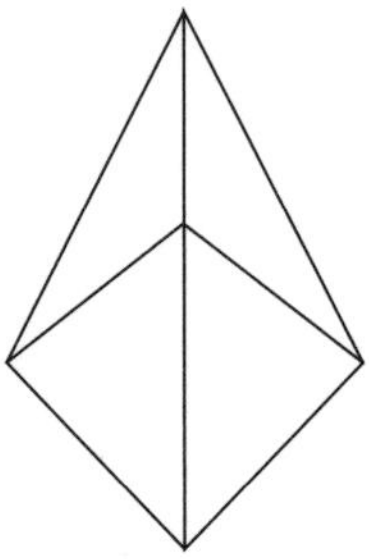

4

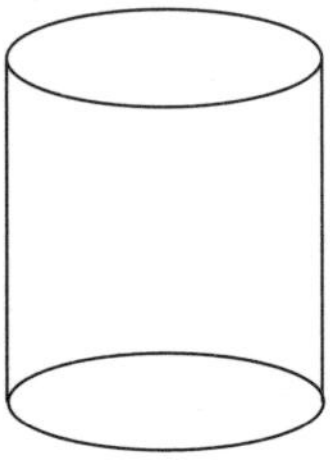

Answers on page 209

5

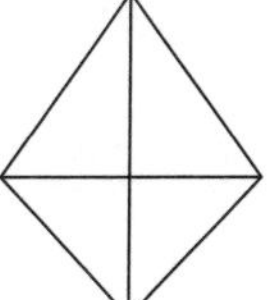

6

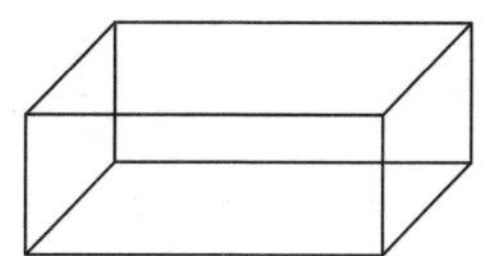

7

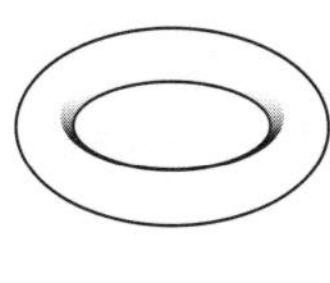

8

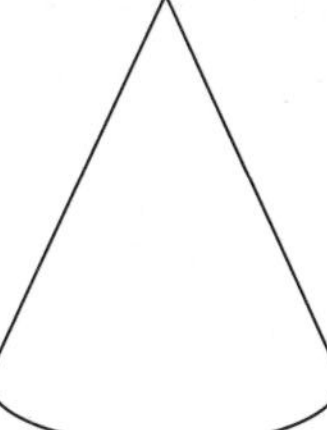

9

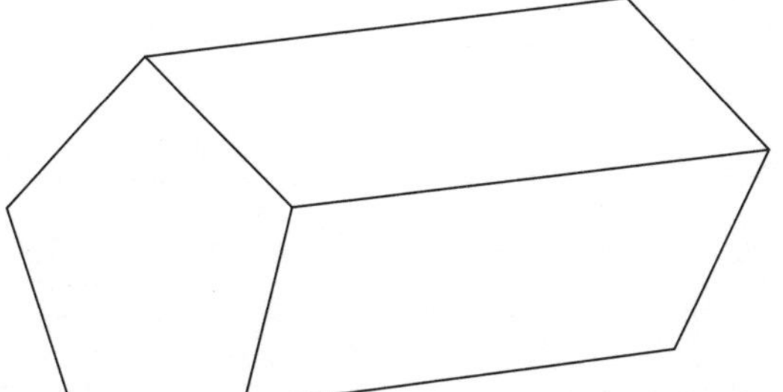

10

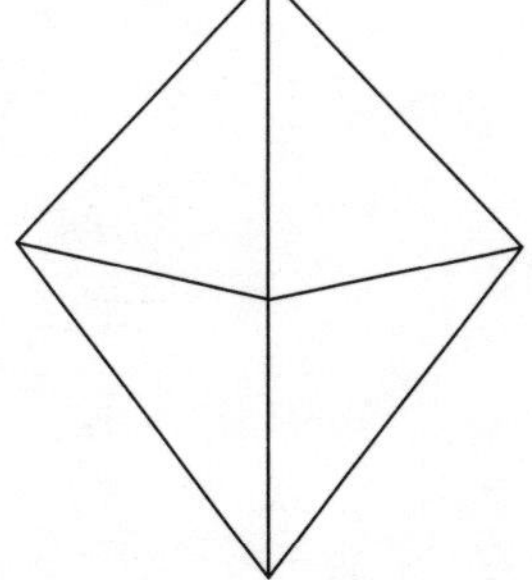

Answers on page 209

Pump up the Volume

To work out how much space any given solid takes up, we need to calculate its volume. When we were looking at area, we talked of units of measurement squared (for instance, cm^2) but since we are now adding a further dimension, we are dealing in units of measurement cubed (e.g. cm^3).

Here are some useful equations for calculating the volume of different shapes (V = Volume; L = Length; H = Height; W = Width; A = Area; R = Radius):

- *V of a cube* = L^3
- *V of a cuboid* = $L \times W \times H$
- *V of a prism* = *A of starting polygon* x *H*
- *V of a cylinder* = $\pi R^2 H$
- *V of a pyramid* = $\frac{A \text{ of Base} \times H}{3}$
- *V of a cone* = = $\frac{\pi R^2 H}{3}$
- *V of a sphere* = $\frac{4\pi R^3}{3}$

Quiz 90

Use the equations on page 138 to work out the volume of each of the solids below (giving your answers to no more than two decimal places):

1

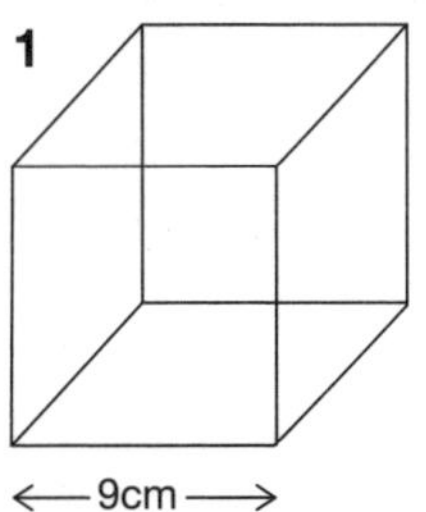

2

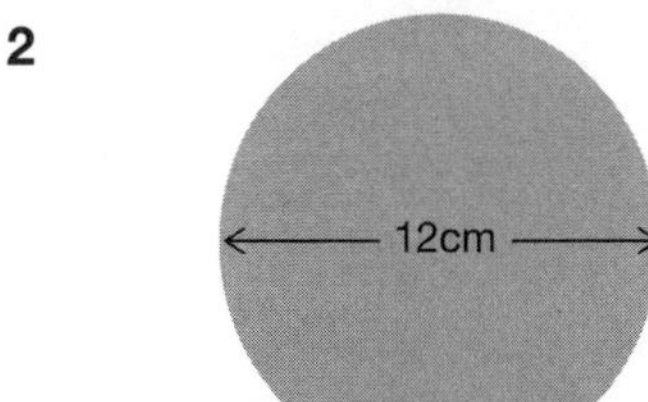

3

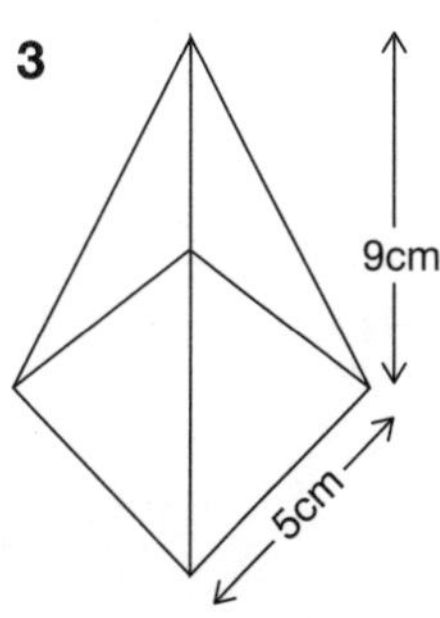

4

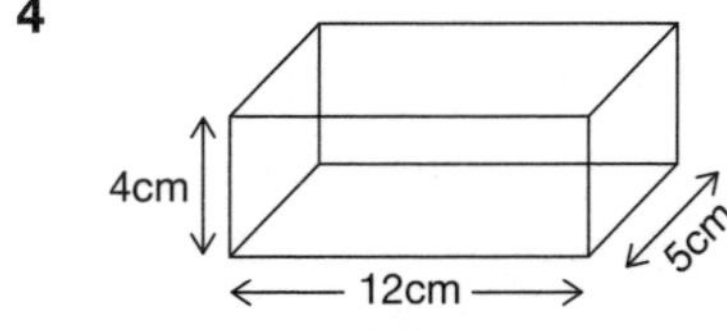

5

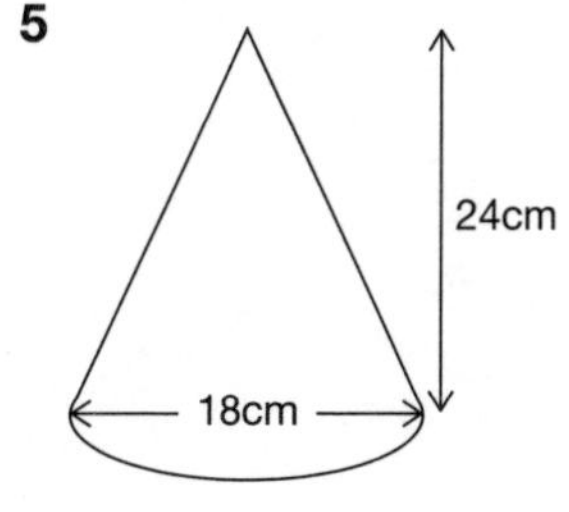

6

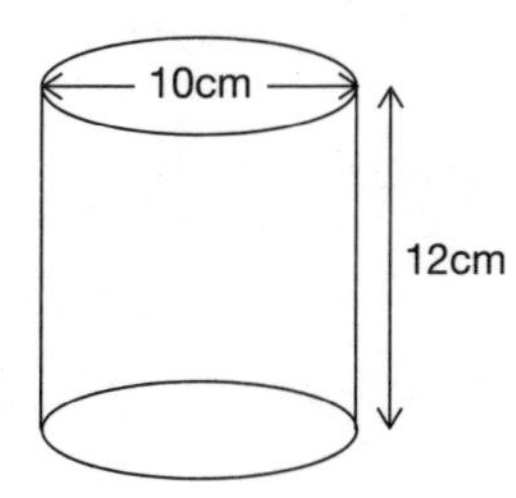

Answers on page 210

Back of the Net

A net is a pattern that may be cut and folded to form a solid. Have a look at the following nets and try to work out which solids they form.

Quiz 91

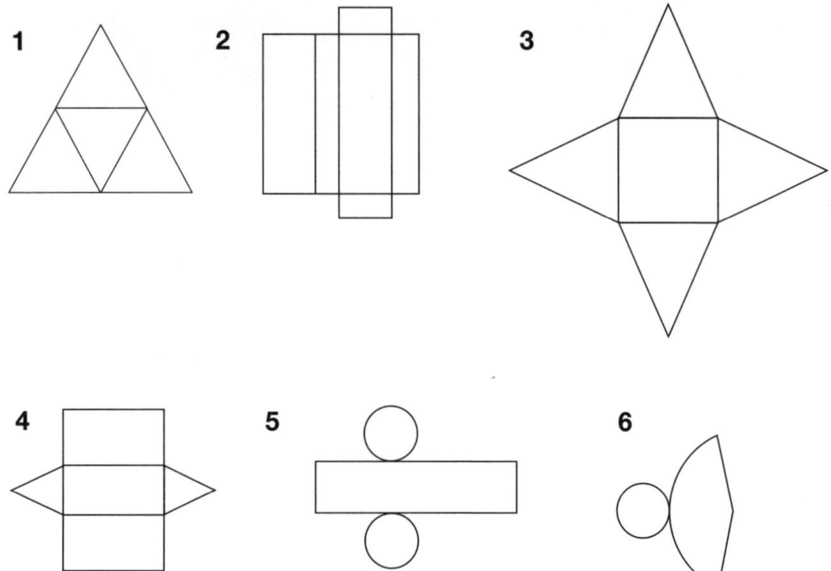

Answers on page 210

Quiz 92

The humble cube has 11 distinct nets. See how many of them you can come up with.

Answers on page 211

Answers on page 211

Answers on page 211

Trigs of the Trade

Trigonometry (or trig, for short) is a largely theoretical branch of mathematics concerned with calculating the angles and lengths of right-angled triangles. In the past it helped sailors to navigate their way across the world's oceans and today it is still of vital importance to engineers, architects and their like. As long as you know the length of one side plus one of the non-right angles, or alternatively the lengths of two sides, you can figure out all of the lengths and all of the angles.

To get to grips with trig, you need to be aware of its special lingo. Firstly, there are the names of the sides of the triangle. The hypotenuse we know from Pythagoras. The opposite side is that opposite a given non-right angle. The adjacent side is the side adjacent to a given non-right angle. Below are two diagrams to show all these sides, though note how the opposite and adjacent sides are dependent on what your given non-right angle is.

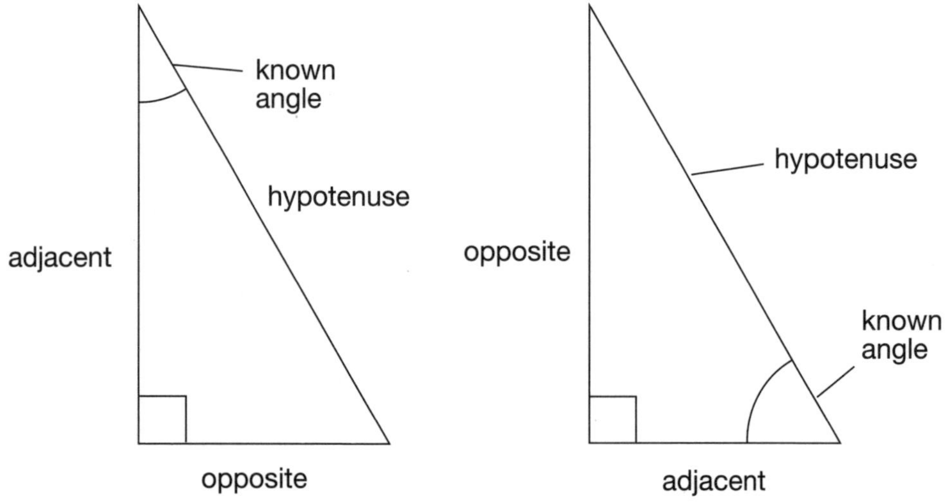

Trigonometry ultimately depends on the fact that there are definite relationships between the angles in a triangle and the ratios of their sides. These three ratios are the sine ratio, cosine ratio and tangent ratio (usually abbreviated to sin, cos and tan). Here are the formulae (and the θ symbol, called theta and from Ancient Greek, is used as shorthand for an angle):

- $\text{Sin } \theta = \dfrac{\text{opposite}}{\text{hypotenuse}}$
- $\text{Cos } \theta = \dfrac{\text{adjacent}}{\text{hypotenuse}}$
- $\text{Tan } \theta = \dfrac{\text{opposite}}{\text{adjacent}}$

Fortunately, most modern calculators have trigonometric functions so you simply need to type in your number and press the correct button (cos, tan or sin) to get an accurate result. If you have the figure for sin, cos or tan but want to know the angle, use the inverse cos, tan and sin functions on your calculator.

Quiz 93

See how you get on with the following questions. In the first three questions, you need to find the marked missing length. In the last two questions, calculate the missing marked angles. State your answers to questions 1 to 3 to a maximum of 2 decimal places. Answer questions 4 and 5 to the nearest degree.

1

a

10cm

40°

2

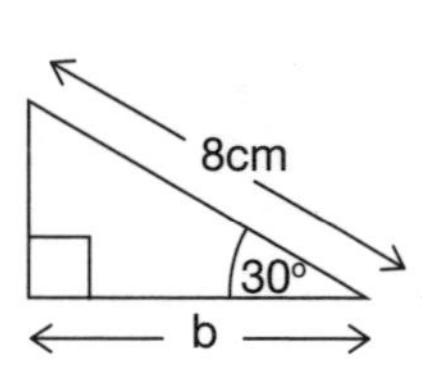

3

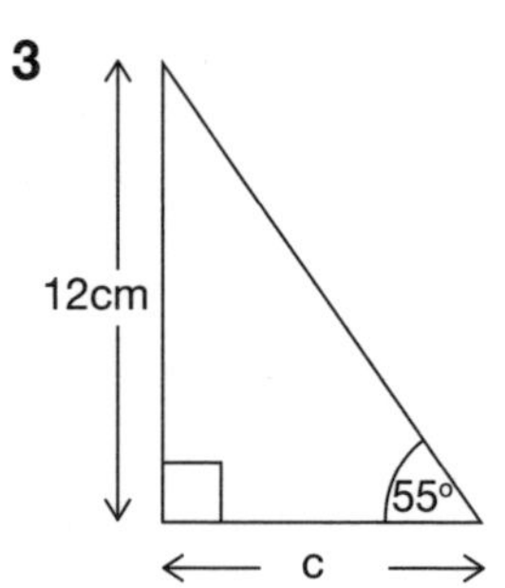

4

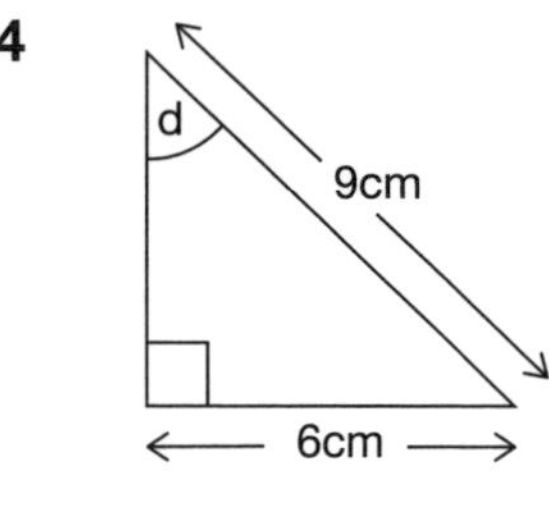

5

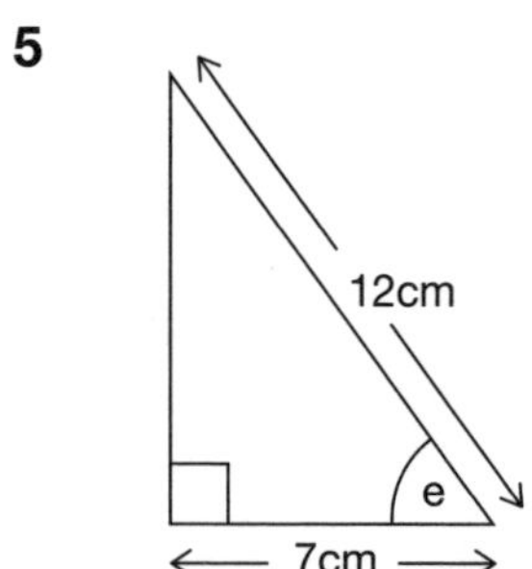

Answers on page 212

A Perfect Match

If you can fold a shape down the middle so that one half covers the other perfectly, the shape may be said to be symmetrical. The line along which it can be folded is known as the axis of symmetry, and some shapes can have several such axes. Have a look at the shapes below and work out how many axes of symmetry each one has. Draw them on to each shape if it helps.

Quiz 94

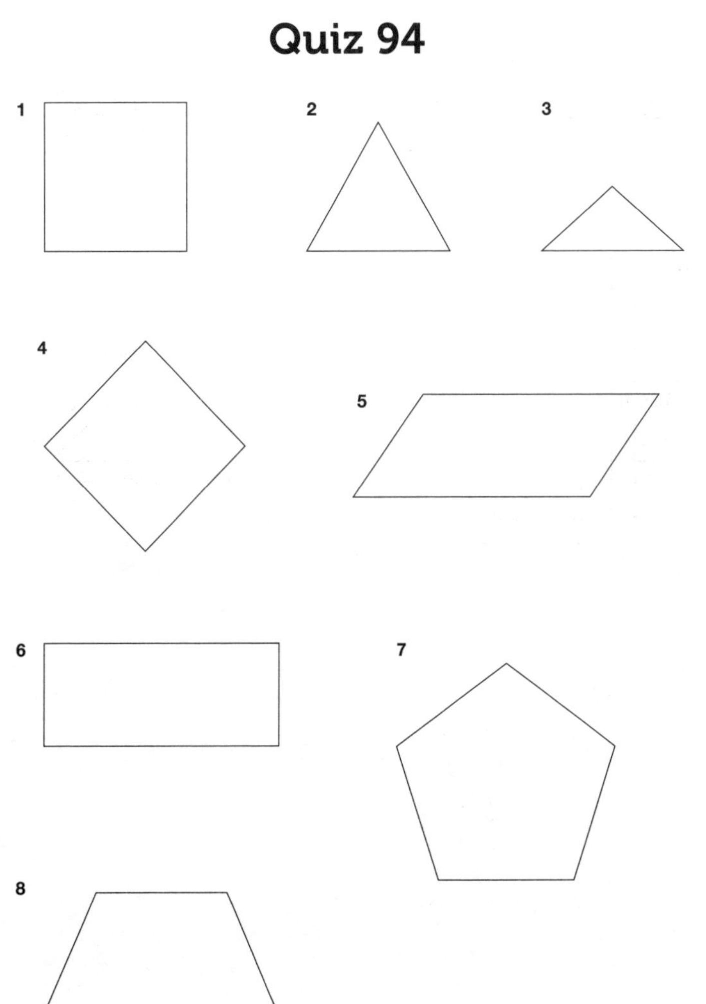

Answers on page 213

Transforming the Situation

In trigonometry, there are three major methods of 'transforming' a shape:

Reflection – where a shape is exactly repeated as a 'mirror image' along a central 'mirror line'

Rotation – where a shape is turned around a central point

Translation – where a shape is reproduced exactly but in a different place

If one shape can be turned into another by any of these transformations, the two shapes are known as congruent. In all other respects (e.g. side length, area, angles and size) the shapes are the same.

Quiz 95

Take a look at the diagrams below. In each case, shape B is a transformation of shape A. Simply tick whether the shape has been reflected, rotated or translated.

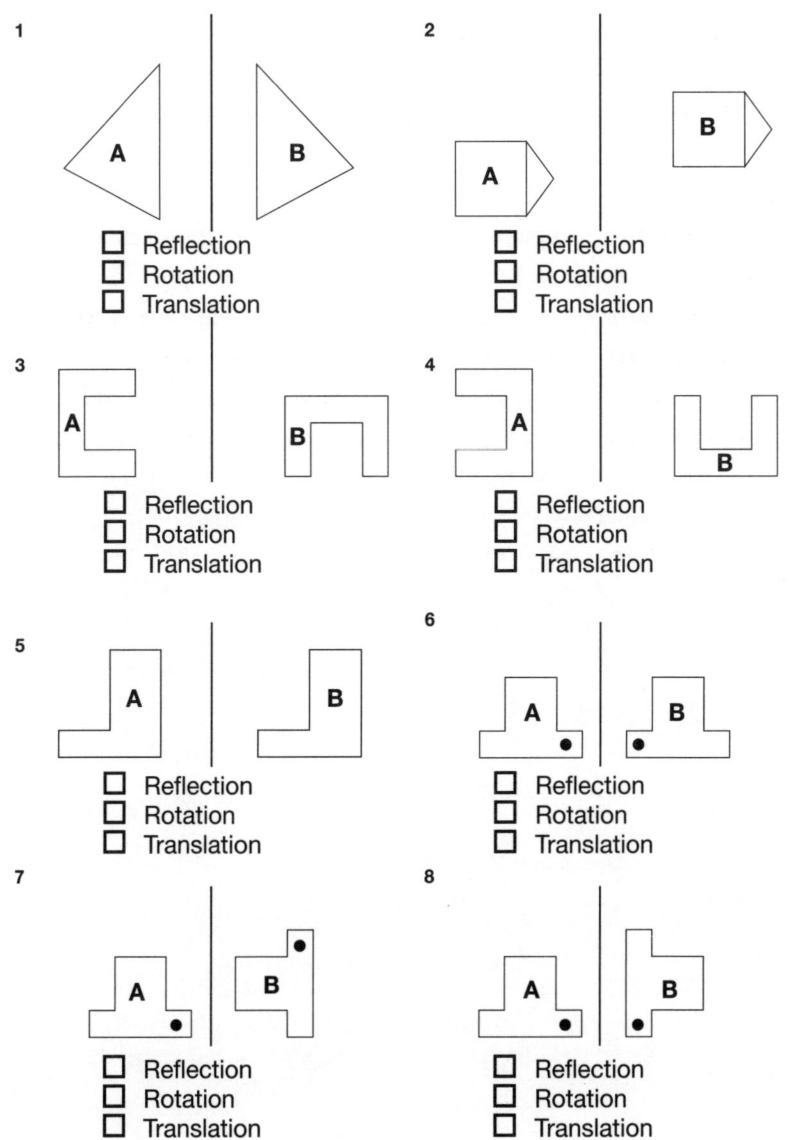

Answers on page 214

Quiz 96: Geometry and Trigonometry Crammer

1) If you attach a triangle to the side of a square, what new shape have you created?
2) A rectangle has a length of 16cm and a width of three quarters that figure. What is its perimeter?
3) Take the same rectangle and calculate its area.
4) A right-angled triangle has a hypotenuse of 13cm. The square on one of the other lengths is 25cm^2. What is the length of the other side?
5) A circle has a radius of 6.5cm. What is its circumference (to 2 decimal places)?
6) A cuboid has a length of 20cm, a width of 16cm and a height of 15cm. What is its volume?
7) A cylinder has a diameter of 12cm and a height of 11cm. What is the volume of the cylinder to the nearest centimetre^3?
8) In a right-angled triangle, the angle a is unknown. If the hypotenuse measures 10cm and the side adjacent to a is 8cm, what is a to the nearest degree?

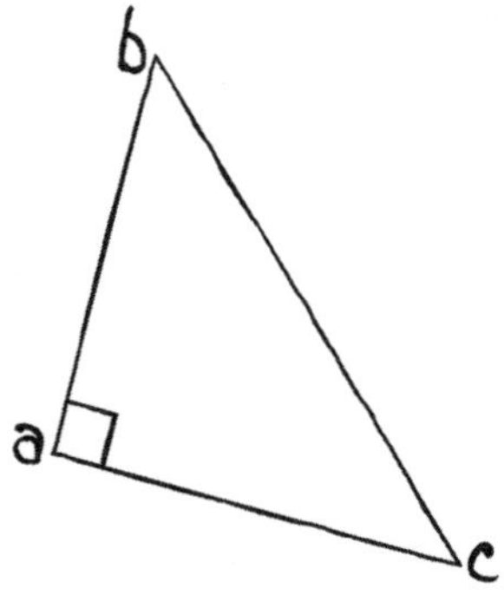

Answers on page 214

THE SUM OF YOUR KNOWLEDGE: PROBLEM SOLVING

In this final section of quizzes we test everything you have learned so far. Now that you have acquired and practised all the principal skills, you can apply them to everyday problems.

Here is a chance for you to evaluate the sum of your knowledge and prove that you've become a maths whiz.

A-Mazing Maths

Quiz 97

Here are a number of maths mazes for you to navigate.

1) By following the arrows, there are 10 paths to take you from start to finish. What is the sum of each path?

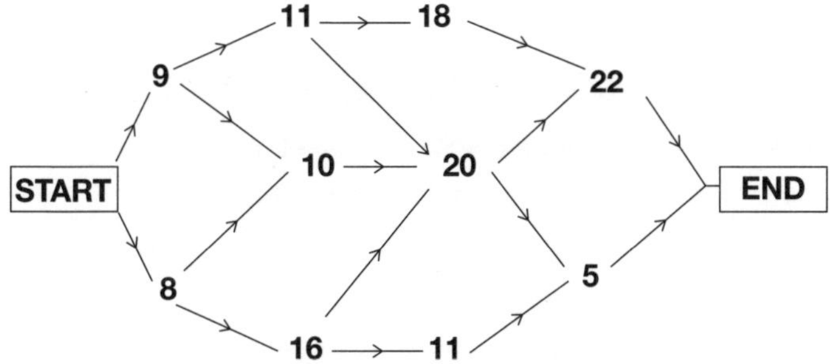

2) Here are two paths to follow. By adding the numbers along the way, you arrive at the end, where the total is 55. On each route is a missing number. Calculate the values of a and b.

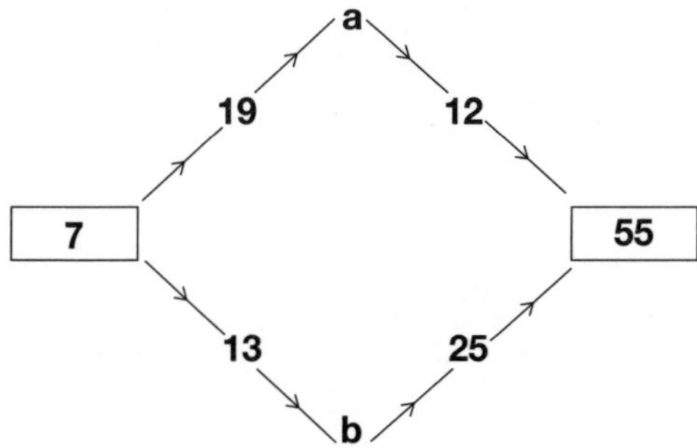

Answers on page 215

3) In this diagram you must subtract the numbers along each route to arrive at your destination, 12. Again, two figures are missing. Give the values of c and d.

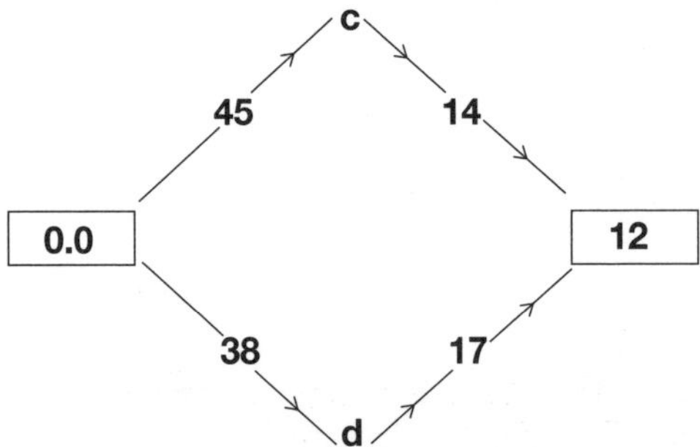

4) As in question 1, there are ten routes to take you from the start (a value of 1) to the end, but this time you must multiply the figures along the route. What are the ten totals?

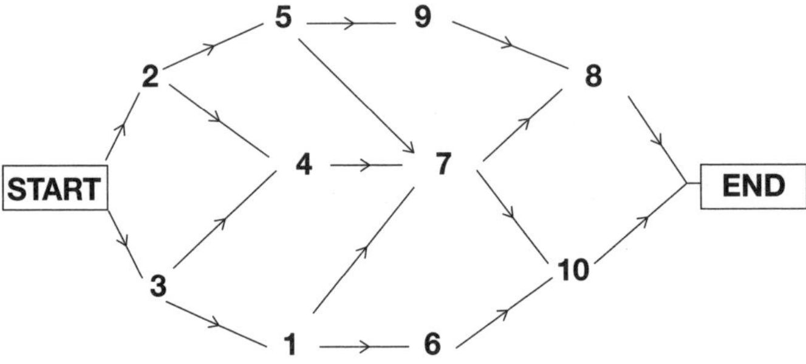

Answers on page 215

5) Here are three possible routes to take you from the start (180) to the end (5). Using the function indicated at each step, only one route is mathematically accurate. Which is it?

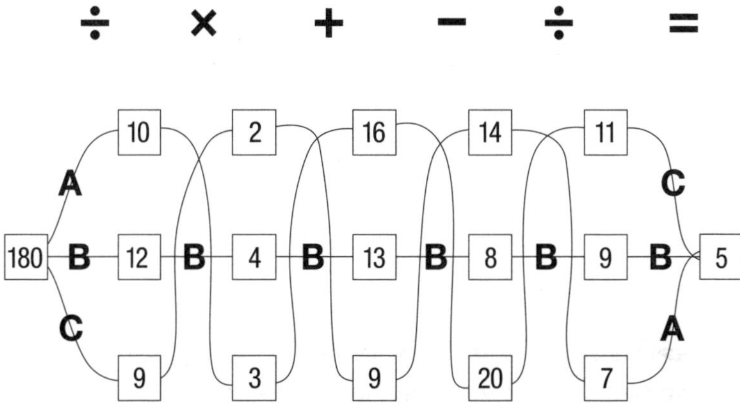

Answers on page 215

A Word to the Wise

Here is a set of word problems that require mathematical reasoning in a practical context. See how many you can get right.

Quiz 98

1) Ben has collected £137 for his school's minibus appeal. Chloe has raised £96 and Jamaal has raised £347. How much have they collected in total?
2) Alf can type at 49 words per minute. He has written an essay longhand that is 4,300 words. Assuming he keeps a constant pace, how much time should he allow himself to type it up (to the nearest minute)?
3) Joe is planning to drive from Paris to Madrid and back again. The distance between Paris and Madrid is about 1,050km, and his car can travel 275km on a full tank of petrol. Assuming he sets off with a full tank, how many times should he need to refuel?
4) The average height of the children in a class is 141 centimetres. If Lisa, the tallest child, is 24 centimetres above the average and her friend Melanie is 80% of her height, how tall is Melanie?
5) Ali wants to record some music from his computer on to compact discs. Each disc can store 74 minutes of music. Ali has 315 songs of average length $3\frac{1}{2}$ minutes that he wants to transfer to disc. How many CDs will he need?

Answers on page 215

6) The English cricket team scored 427 in their first innings but 146 runs less in the second innings. If they beat Australia by 3 clear runs, how much was the total of the Australian innings?
7) 72 people were asked to name their favourite season of the year. An eighth said winter and a third said spring. If 14 answered autumn, how many chose summer?
8) Giovanni the Ice Cream Man chops up 1 chocolate bar to provide enough flakes for 3 '99s'. On a particularly hot day, he sells 216 '99s'. If there are 12 chocolate bars to a box, how many new boxes will he need to order at the end of the day?

Answers on page 215

Number Crunching

In this section we test your combined mathematical and crossword skills.

Quiz 99

Use the clues below to complete the crossword.

Across

1. 6 x 22
3. Convert 1100 from binary.
5. What is the next prime number after 41?
6. If two angles in a triangle are 30° and 42°, what is the value of the final angle?
7. Give the cube root of 64.
8. a = 13 x 14 –41. What is a worth?
9. Multiply 3 by 3^2.

Answers on page 216

11. A pie is cut into equal slices of 30° each.
    How many slices are there?
12. 384 – 228
14. 13 x 51
15. If 2b + 12 = 3b – 12, what is the value of b?

Down

1. How much is a dozen dozen?
2. 99 – 66
3. Express 5 as a binary number.
4. If 3c = 84, what is c worth?
6. Express $^{84}/_{6}$ as a whole number.
8. What is the diameter (to the nearest whole cm) of a circle of circumference 53.4cm?
9. A rectangle has sides of 8cm by 28.25cm.
   What is its area in cm^2?
10. What is IV x CXLI expressed as an Arabic number?
11. 208 ÷ 13
13. Four friends plan to drive in stages from London to York (a distance of 283km). If Paul takes the first leg (67km), Jake drives the next (74km) and Tony the third (90km), how far will Keith have to drive on the final leg?

Answers on page 216

The Big One

Here is a final quiz to test just how much you know. Prepare for brain-strain!

Quiz 100

1) What is 11 x 12?

2) A sandwich shop sells 37 cheese and tomato sandwiches, 42 ham and pickle and 57 tuna mayonnaise. How many sandwiches does it sell in total?

3) –12 x –15 =

4) Here's a sum in binary. Solve it, giving your answer in binary: 101 x 100000 =

5) A library contained 2,387 books but it was running out of shelf space so the librarian decided to sell 446 titles. Alas, a short while later a fire broke out and another 294 titles perished. A further 183 books had to be withdrawn when the local river swelled and flooded the library. Soon afterwards another 4 books were stolen and 1 was chewed by a stray dog. How many books did the unfortunate librarian have left after all of these events?

6) What is 880 divided by 32?

7) 60 people were asked which of the 20 football teams in the English Premier League they supported, with the results expressed in a pie chart. If 7 people chose Everton, what would be the size in degrees of Everton's slice of the pie?

Answers on page 217

8) Calculate $\frac{3}{8} \div \frac{5}{6}$.

9) Mordechai is hooked on fashion and spends £77 per day on clothes for the whole of the month of May. How much does he spend in total?

10) What is the cube root of 2,197?

11) The square on the hypotenuse of a right-angled triangle equals 289cm^2. If one of the other lengths of the triangle is 8cm, how long is the missing length?

12) Solve the following Roman numerals sum, giving your answer in Roman numerals too. CLXXIX – LXXXVII =

13) Look at this number sequence and deduce the next two numbers: 3, 4, 6, 10, 18, 34, ?, ?

14) Convert $\frac{437}{38}$ into a mixed number.

15) If 3(a + 5) = 36, what does a equal?

16) If you add together 7% plus $\frac{3}{5}$, what do you get (expressing your answer as a decimal)?

17) A circle has a circumference of 138.2cm. What is its radius, to the nearest cm?

18) Sheila has seen a sporty little car that she wants to buy. It normally costs £14,750 but for one day only there is a 22% reduction. What is the new price?

19) A game of snooker begins with 15 red balls, 1 white one, 1 yellow, 1 green, 1 brown, 1 blue, 1 pink and 1 black. If all the balls are put into a bag, what are your chances of randomly picking out 2 balls that are neither red nor white?

Answers on page 217

20) Lovelorn Hernando has plotted the length in days of each of his 7 most recent relationships. Give the mean, median and mode length of his relationships to the nearest whole number: 6, 14, 14, 59, 11, 234, 738

21) What is 4.6 x 9.5?

22) If the current exchange rate is $1.00 to £0.63 and I have $420 to change to pounds, how much sterling will I get?

23) If 3m + 2n = 75 and 4m – n = 67, what are the values of m and n?

24) If a ladder 8 metres long stands against a wall at an angle of 35° to the ground, how high up the wall will the end of the ladder be (to the nearest centimetre)?

25) I am thinking of a prime number composed of two digits, the product of which totals 24. What number am I thinking of?

Answers on page 217

ANSWERS

Quiz 1

1)	+	m)	plus
2)	–	g)	minus
3)	×	k)	multiplied by
4)	÷	h)	divided by
5)	=	j)	equals
6)	≠	c)	does not equal
7)	≈	l)	is approximately equal to
8)	<	i)	is less than
9)	>	f)	is greater than
10)	≤	b)	is less than or equal to
11)	≥	n)	is greater than or equal to
12)	√	e)	square root
13)	∞	a)	infinity
14)	π	d)	pi (used in numerous calculations related to circles)

Quiz 2

1) 16
2) 8
3) 24
4) 26
5) 7
6) 40
7) 19
8) 44
9) 56
10) 5

Quiz 3

1	2	3	4	5	6	7	8	9	10	11	12
2	4	6	8	10	12	14	16	18	20	22	24
3	6	9	12	15	18	21	24	27	30	33	36
4	8	12	16	20	24	28	32	36	40	44	48
5	10	15	20	25	30	35	40	45	50	55	60
6	12	18	24	30	36	42	48	54	60	66	72
7	14	21	28	35	42	49	56	63	70	77	84
8	16	24	32	40	48	56	64	72	80	88	96
9	18	27	36	45	54	63	72	81	90	99	108
10	20	30	40	50	60	70	80	90	100	110	120
11	22	33	44	55	66	77	88	99	110	121	132
12	24	36	48	60	72	84	96	108	120	132	144

Quiz 4

1) 2
2) –11
3) 4
4) –5
5) –12
6) –4
7) –3
8) 9
9) –26
10) 5

Quiz 5

1) –12
2) 42
3) –19
4) –84
5) 9
6) 840
7) –6
8) –133
9) 630
10) –6

Quiz 6

1) 6
2) 4
3) 7
4) 7
5) 2700
6) 362,750
7) 28,750
8) 0.013825

Quiz 7

1) 131 cards

$$\begin{array}{cccc} & {\scriptstyle 1} & {\scriptstyle 1} & \\ & & 4 & 3 \\ + & & 8 & 8 \\ \hline & 1 & 3 & 1 \end{array}$$

2) 160 hours

$$\begin{array}{cccc} & {\scriptstyle 1} & {\scriptstyle 2} & \\ & & 3 & 6 \\ & & 4 & 1 \\ & & 4 & 8 \\ + & & 3 & 5 \\ \hline & 1 & 6 & 0 \end{array}$$

3) 1,617 pupils

$$\begin{array}{ccccc} & & {\scriptstyle 1} & {\scriptstyle 1} & \\ & & 6 & 6 & 2 \\ + & & 9 & 9 & 5 \\ \hline & 1 & 6 & 1 & 7 \end{array}$$

4) 2,577 residents

$$
\begin{array}{rrrrl}
{}^{1} & & & & \\
 & 5 & 5 & 2 & + \\
 & 8 & 1 & 1 & \\
1 & 2 & 1 & 4 & \\
\hline
2 & 5 & 7 & 7 &
\end{array}
$$

5) £8,892

$$
\begin{array}{rrrrl}
{}^{1} & {}^{1} & {}^{1} & & \\
1 & 3 & 5 & 2 & + \\
2 & 8 & 8 & 9 & \\
4 & 6 & 5 & 1 & \\
\hline
8 & 8 & 9 & 2 &
\end{array}
$$

Quiz 8

1) 29 cards

$$
\begin{array}{rrl}
\not{9}8 & {}^{1}7 & - \\
6 & 8 & \\
\hline
2 & 9 &
\end{array}
$$

2) 38 hours

$$
\begin{array}{rrl}
\not{1}0\not{3}\,{}^{1}2 & {}^{1}7 & - \\
9 & 9 & \\
\hline
3 & 8 &
\end{array}
$$

3) 410 seats

$$
\begin{array}{rrrrl}
\not{2}1 & {}^{1}0 & 2 & 7 & - \\
1 & 6 & 1 & 7 & \\
\hline
 & 4 & 1 & 0 &
\end{array}
$$

4) 1,766 residents

$$
\begin{array}{rrrrl}
\not{2}1 & {}^{1}5 & 7 & 7 & - \\
 & 8 & 1 & 1 & \\
\hline
1 & 7 & 6 & 6 &
\end{array}
$$

5) £4,919

$$
\begin{array}{rrrrl}
\not{8}7 & {}^{1}8 & \not{9}8 & {}^{1}2 & - \\
3 & 9 & 7 & 3 & \\
\hline
4 & 9 & 1 & 9 &
\end{array}
$$

Quiz 9

1) 136

	8	
10	10 x 8 = 80	
7	7 x 8 = 56	(80 + 56 = 136)

2) 208

	10	3	
10	10 x 10 = 100	10 x 3 = 30	
6	6 x 10 = 60	6 x 3 = 18	(100 + 60 + 30 + 18 = 208)

3) 510

	10	5	
30	30 x 10 = 300	30 x 5 = 150	
4	4 x 10 = 40	4 x 5 = 20	(300 + 40 + 150 + 20 = 510)

4) 1,092

	20	6	
40	40 x 20 = 800	40 x 6 = 240	
2	2 x 20 = 40	2 x 6 = 12	(800 + 40 + 240 + 12 = 1,092)

5) 9,204

	50	2	
100	100 x 50 = 5000	100 x 2 = 200	
70	70 x 50 = 3500	70 x 2 = 140	
7	7 x 50 = 350	7 x 2 = 14	(5,000 + 3,500 + 350 + 200 + 140 + 14 = 9,204)

Quiz 10

1) £832

52 x 16

	¹5	2 ×
	1	6
3	1	2
5	2	0
8	3	2

2) 2,057 stamps

11 x 121

	¹1	2	1 ×
		1	7
	8	4	7
1	2	1	0
2	0	5	7

3) £4,675

187 x 25

```
  1 1
  4 3
  1 8 7  ×
    2 5
---------
  9 3 5
3 7 4 0
---------
4 6 7 5
```

4) £18,506

487 x 38

```
    2 2
    6 5
    4 8 7  ×
      3 8
-----------
  3 8 9 6
1 4 6 1 0
-----------
1 8 5 0 6
```

5) 8,760 hours

365 x 24

```
  1 1
  2 2
  3 6 5  ×
    2 4
---------
1 4 6 0
7 3 0 0
---------
8 7 6 0
```

6) 86,400 seconds

60 x 60 x 24

$$\begin{array}{r} 6\ 0 \\ 6\ 0 \\ \hline 0\ 0 \\ 3\ 6\ 0\ 0 \end{array} \times$$

$$\begin{array}{r} 3\ 6\ 0\ 0 \\ 2\ 4 \\ \hline 8\ 6\ 4\ 0\ 0\ 0 \end{array} \times$$

Quiz 11

1) 12

$$\begin{array}{r|l} & 1\ 2 \\ \cline{2-2} 7 & 8\ ^{1}4 \end{array}$$

2) 16

$$\begin{array}{r|l} & 0\ 1\ 6 \\ \cline{2-2} 12 & 1\ ^{1}9\ ^{7}2 \end{array}$$

3) £27.50

$$\begin{array}{r|l} & 0\ \ 2\ 7.\ 5 \\ \cline{2-2} 10 & 2\ ^{2}7\ ^{7}5.^{5}0 \end{array}$$

4) 22g

$$\begin{array}{r|l} & 0\ 2\ 2 \\ \cline{2-2} 9 & 1\ ^{1}9\ ^{1}8 \end{array}$$

5) 29

$$\begin{array}{l} \quad 0\ 2\ 9 \\ 6\overline{\smash{)}1\ {}^{1}7\ {}^{5}4} \end{array}$$

6) 37.5 metres

$$\begin{array}{l} \quad 0\ 3\ 7\ .\ 5 \\ 8\overline{\smash{)}3\ {}^{3}0\ {}^{6}0\ .\ {}^{4}0} \end{array}$$

7) $2.\dot{6}$

$$\begin{array}{l} \quad 2\ .\ 6\ 6\ 6\ 6 \\ 3\overline{\smash{)}8\ .\ {}^{2}0\ {}^{2}0\ {}^{2}0\ {}^{2}0} \end{array}$$

Quiz 12

1) 13

$$\begin{array}{r} 13 \\ 12\overline{\smash{)}156} \\ 12 \\ \hline 36 \end{array}$$

2) 27

$$\begin{array}{r} 27 \\ 31\overline{\smash{)}837} \\ 62 \\ \hline 217 \end{array}$$

3) 123

```
      123
87 | 10701
     87
   -------
     200
    -174
   -------
       261
```

Quiz 13

1) 1 and 2
2) 1, 2 and 4
3) 1 and 7
4) 1, 3 and 9
5) 1 and 11
6) 1, 3, 7 and 21

Therefore, 2, 7 and 11 are prime.

Quiz 14

1) 17
2) 37
3) 83
4) 419
5) 733

Quiz 15

1) 13
2) 29
3) 89
4) 421
5) 997

Quiz 16

1) 3:4
2) 156 girls
3) 150 seedlings
4) 15 eggs
5) 27ml
6) 78 euros

Quiz 17

1) 1
2) 1000
3) 100
4) 50
5) 5
6) 10
7) 500

Quiz 18

1) VII (2 + 5 = 7)
2) V (8 – 3 = 5)
3) XXXII (13 + 19 = 32)
4) XXIII (46 – 23 = 23)
5) CCXLVIII (122 + 126 = 248)
6) CL (312 – 162 = 150)
7) CD (148 + 252 = 400)
8) DCCCLXXVII (1822 – 945 = 877)

Quiz 19

1) 4
2) 5
3) 7
4) 16
5) 35

Quiz 20

1) 1001
2) 1011
3) 1110
4) 11001
5) 1100100

Quiz 21

1) 0.5
2) 0.75
3) 2.25
4) 12.125
5) 4.875

Quiz 22

1) 1100 (8 + 4 = 12)
2) 1111 (17 – 2 = 15)
3) 110001 (7 x 7 = 49)
4) 1011010 (64 + 32 – 6 = 90)
5) 10.1 (1.25 + 1.25 = 2.5)
6) 1111.101 (5 x 3.125 = 15.625)

Quiz 23

Base number (A)	**A^2**	**A^3**
2	4	8
3	9	27
4	16	64
5	25	125
6	36	216
7	49	343
8	64	512
9	81	729
10	100	1,000
11	121	1,331
12	144	1,728

Quiz 24

1) 256
2) 3,125
3) 46,656
4) 823,543
5) 16,777,216
6) 387,420,489
7) 10,000,000,000

Quiz 25

1) 9
2) 13
3) 5
4) 16
5) 10
6) 100
7) 20
8) 30
9) 70
10) 14

Quiz 26

1) $9 + (6 \div 2) = 12$
2) $7 - (3 + 8) = -4$
3) $(112 \div 2) \times 7 + 2 = 394$
4) $5 \times (18 + 6) - 10 = 110$
5) $(9 \times 7) \times (10 + 2) = 756$
6) $(4^2 + 8) \times (15 \div 5) - 12 = 60$

Quiz 27

1) 28
2) 5
3) 73
4) 113
5) 20

Quiz 28

1) 64, 128. A simple sequence in which each new number is double the previous one.
2) 13, 17. The first seven primary numbers.
3) 95, 191. If n is the previous number in the sequence, each new number is 2n + 1.
4) 34, 55. An extract from the famous Fibonacci sequence, in which each number is equal to the sum of the previous two numbers in the sequence.
5) 1100, 1110. Even numbers (starting from 2) written in binary.
6) 244, 730. If n is the previous number in the sequence, each new number is 3n – 2.

Quiz 29

1) 81
2) 6
3) 817
4) 44
5) 53
6) 875g
7) 68
8) 32

Quiz 30

1) $\frac{1}{4}$
2) $\frac{3}{4}$
3) $\frac{2}{3}$
4) $\frac{5}{8}$
5) $\frac{9}{10}$

Quiz 31

1) $\frac{3}{4}$
2) $\frac{9}{10}$
3) $\frac{4}{5}$
4) $\frac{1}{3}$
5) $\frac{1}{17}$
6) $\frac{6}{7}$

Quiz 32

1) $^{2}/_{9}$
2) 1 $^{1}/_{5}$
3) $^{9}/_{10}$ ($^{6}/_{10}$ + $^{3}/_{10}$)
4) $^{1}/_{12}$ ($^{9}/_{12}$ – $^{8}/_{12}$)
5) 1 $^{4}/_{63}$ ($^{18}/_{63}$ + $^{49}/_{63}$ => $^{67}/_{63}$)
6) $^{3}/_{56}$ ($^{35}/_{56}$ – $^{32}/_{56}$)

Quiz 33

1) $^{8}/_{9}$
2) 1 $^{3}/_{5}$
3) 1 $^{3}/_{4}$
4) 2 $^{2}/_{11}$
5) 2 $^{2}/_{5}$
6) 3 $^{6}/_{7}$

Quiz 34

1) $^{8}/_{5}$
2) $^{15}/_{4}$
3) $^{12}/_{5}$
4) $^{61}/_{8}$
5) $^{471}/_{100}$
6) 1 $^{1}/_{8}$
7) 2 $^{3}/_{16}$
8) 7 $^{4}/_{7}$
9) 11 $^{11}/_{12}$
10) 3 $^{6}/_{17}$

Quiz 35

1) True
2) True
3) True
4) False
5) False
6) True

Quiz 36

1) $^{5}/_{57}$ (Cancel down to $^{1}/_{3}$ x $^{5}/_{19}$)
2) $^{29}/_{77}$ (Cancel down to $^{29}/_{7}$ x $^{1}/_{11}$)
3) $^{53}/_{63}$ (Cancel down to $^{53}/_{7}$ x $^{1}/_{9}$)
4) $^{3}/_{4}$ (Cancel down to 3 x $^{1}/_{4}$)
5) $^{22}/_{57}$ (Cancel down to $^{11}/_{19}$ x $^{2}/_{3}$)
6) $^{7}/_{16}$ (Cancel down to $^{42}/_{32}$ x $^{1}/_{3}$ => $^{42}/_{96}$)

Quiz 37

1) $^{2}/_{3}$ ($^{1}/_{2}$ x $^{4}/_{3}$)
2) $^{16}/_{21}$ ($^{2}/_{3}$ x $^{8}/_{7}$)
3) 1 ($^{1}/_{8}$ x $^{8}/_{1}$)
4) 1 $^{17}/_{18}$ ($^{5}/_{6}$ x $^{7}/_{3}$)
5) 1 $^{5}/_{9}$ ($^{7}/_{9}$ x $^{2}/_{1}$)
6) $^{55}/_{63}$ ($^{11}/_{14}$ x $^{10}/_{9}$ => $^{110}/_{126}$)

Quiz 38

1) 5.4

1
1.8 +
3.6
——
5.4

2) 2.7

6~~7~~.^{1}2 –
4 . 5
——
2 . 7

3) 2.295

1
1.875 +
0.420
——
2.295

4) 2.923

5~~6~~.^{1}843 –
3.920
——
2.923

5) 13.993

1
7.500 +
2.870
3.623
——
13.993

6) 2.892

$$
\begin{array}{r}
{}^{8}\not{9}.{}^{17}\not{8}42\ - \\
6.950 \\
\hline
2.892
\end{array}
$$

Quiz 39

1) 8.4
2) 7.38
3) 1.54
4) 1.65
5) 3.24
6) 9.66

Quiz 40

1) 0.64
2) 1.4
3) 2.4
4) 3 (Your sum becomes $^{45}/_{15}$)
5) 22.75 (Your sum becomes $^{182}/_{8}$)
6) $1.\dot{2}5\dot{9}$ (Your sum becomes $^{34}/_{27}$)

Quiz 41

1/50	0.02	2%
1/40	0.025	2.5%
1/25	0.04	4%
1/20	0.05	5%
1/10	0.1	10%
1/8	0.125	12.5%
1/5	0.2	20%
1/4	0.25	25%
1/3	$0.\dot{3}$	33.33%
3/8	0.375	37.5%
2/5	0.4	40%
1/2	0.5	50%
3/5	0.6	60%
5/8	0.625	62.5%
2/3	$0.\dot{6}$	66.66%
7/10	0.7	70%
3/4	0.75	75%
4/5	0.8	80%
7/8	0.875	87.5%
9/10	0.9	90%
1	1.0	100%

Quiz 42

1) Value of reduction £4; new price £36
2) Value of reduction £140; new price £420
3) Value of reduction £21.60; new price £98.40
4) Value of reduction £450; new price £300
5) Value of reduction £2,050; new price £8,200

Quiz 43

1) 80% ($^{8}/_{10}$ x 100)
2) 35% ($^{7}/_{20}$ x 100 ⇨ $^{7}/_{1}$ x 5)
3) 65% ($^{78}/_{120}$ x 100 ⇨ $^{78}/_{6}$ x 5 ⇨ $^{390}/_{6}$
4) 17.5% ($^{28}/_{160}$ x 100 ⇨ $^{28}/_{8}$ x 5 ⇨ $^{140}/_{8}$)
5) 63% ($^{504}/_{800}$ x 100 ⇨ $^{504}/_{8}$)

Quiz 44

1) 62.5% (65 – 40 = 25; 25 ÷ 40 = 0.625; 0.625 x 100 = 62.5)
2) 10.7% (12.2 – 10.9 = 1.3; 1.3 ÷ 12.2 = 0.107; 0.107 x 100 = 10.7)
3) 9.2% (830 – 760 = 70; 70 ÷ 760 = 0.092; 0.092 x 100 = 9.2)
4) 19.4% (67 – 54 = 13; 13 ÷ 67 = 0.194; 0.194 x 100 = 19.4)
5) 11% (1,347 – 1,213 = 134; 134 ÷ 1,213 = 0.11; 0.11 x 100 = 11)

Quiz 45

1) Possible
2) Impossible
3) Impossible
4) Possible
5) Possible
6) Impossible

Quiz 46

1) Uncertain
2) Certain
3) Uncertain
4) Certain
5) Certain
6) Uncertain

Quiz 47

1) $\frac{1}{4}$ ($\frac{1}{2} \times \frac{1}{2}$)
2) $\frac{1}{8}$ ($\frac{1}{2} \times \frac{1}{2} \times \frac{1}{2}$)
3) $\frac{1}{32}$ ($\frac{1}{2} \times \frac{1}{2} \times \frac{1}{2} \times \frac{1}{2} \times \frac{1}{2}$)
4) $\frac{1}{3}$
5) $\frac{1}{9}$ ($\frac{1}{3} \times \frac{1}{3}$)
6) $\frac{1}{36}$ ($\frac{1}{6} \times \frac{1}{6}$)
7) $\frac{1}{46656}$ ($\frac{1}{6} \times \frac{1}{6} \times \frac{1}{6} \times \frac{1}{6} \times \frac{1}{6} \times \frac{1}{6}$)
8) $\frac{1}{52}$
9) $\frac{1}{4}$
10) $\frac{3}{13}$
11) $\frac{5}{442}$ ($\frac{6}{52} \times \frac{5}{51}$)
12) $\frac{1}{270725}$ ($\frac{4}{52} \times \frac{3}{51} \times \frac{2}{50} \times \frac{1}{49}$)

Quiz 48

1) Odds of winning the lottery: 1 in 13,983,816
   $\frac{6}{49} \times \frac{5}{48} \times \frac{4}{47} \times \frac{3}{46} \times \frac{2}{45} \times \frac{1}{44}$
   ⇨ $\frac{720}{10,068,347,500}$
   ⇨ $\frac{1}{13,983,816}$
   (Better not give up the day job just yet.)
2) Odds of being dealt a royal flush: 1 in 649,740
   $\frac{20}{52} \times \frac{4}{51} \times \frac{3}{50} \times \frac{2}{49} \times \frac{1}{48}$
   ⇨ $\frac{480}{311,187,520}$
   ⇨ $\frac{1}{649,740}$

Quiz 49

1) $^{17}/_{20}$
2) $^{35}/_{54}$
3) 23
4) $^{17}/_{26}$
5) $2\ ^{2}/_{7}$
6) £24.50
7) 41%
8) $^{1}/_{18}$

Quiz 50

1) Quantitative
2) Quantitative
3) Qualitative
4) Quantitative
5) Qualitative
6) Qualitative

Quiz 51

1) Continuous
2) Discrete
3) Discrete
4) Continuous
5) Discrete
6) Continuous

Quiz 52

1) Mean = 179.4cm; Median = 176cm; Mode = No mode
2) Mean = £8.19; Median = £7.45; Mode = £7.45
3) Mean = 36.71 years; Median = 33 years; Mode = 20 years
4) Mean = 115.9 mins; Median = 116 mins; Mode = 116 mins
5) Mean = 45.9%; Median = 49.5%; Mode = 53%

Quiz 53

1) Football
2) Angling
3) 18 people
4) 37 people
5) 17 people

Quiz 54

1) 9° (360° ÷ 40 cars)
2) Yellow
3) 11
4) Black and blue (7 of each)
5) 13

Quiz 55

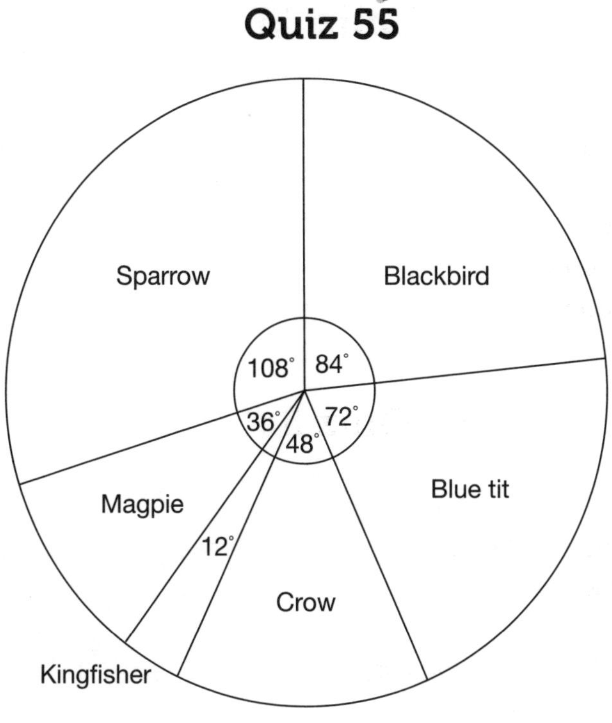
Sparrow
Blackbird
108°
84°
36°
72°
48°
Magpie
12°
Blue tit
Crow
Kingfisher

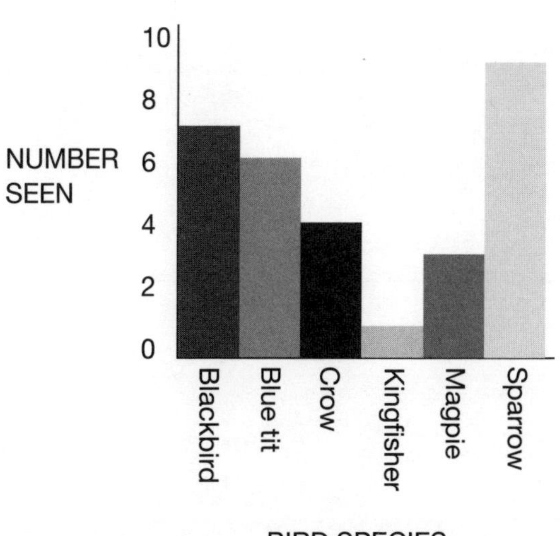
10
8
NUMBER SEEN
6
4
2
0
Blackbird
Blue tit
Crow
Kingfisher
Magpie
Sparrow
BIRD SPECIES

Quiz 56

1) Negative correlation
2) No correlation
3) Positive correlation
4) No correlation
5) Positive correlation
6) Negative correlation

Quiz 57

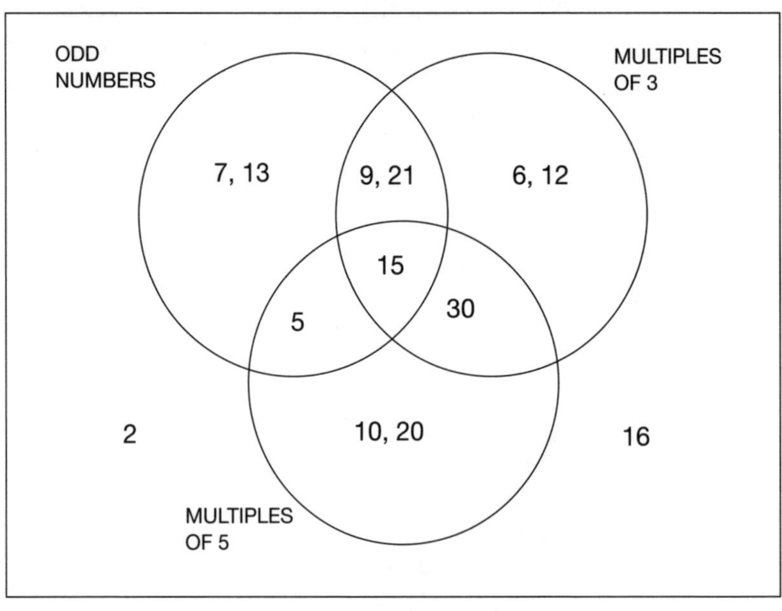

Quiz 58

Milli (m)	1 thousandth
Mega (M)	1 million
Hecto (h)	1 hundred
Nano (n)	1 billionth
Giga (G)	1 billion
Centi (c)	1 hundredth
Tera (T)	1 trillion
Micro (μ)	1 millionth
Kilo (k)	1 thousand

Quiz 59

1 kilogram	2.2 pounds
1 tonne	0.98 tons
1 centimetre	0.39 inches
1 metre	3 feet and 3.4 inches
1 kilometre	0.62 miles
1 square metre	1.19 square yards
1 hectare	2.47 acres
1 litre	1.76 pints

Quiz 60

1 ounce	28 grams
1 pound	0.454 kilograms
1 stone	6.35 kilograms
1 ton	1016 kilograms
1 inch	2.54 centimetres
1 foot	0.305 metres
1 yard	91.4 centimetres
1 mile	1.609 kilometres
1 acre	0.4 hectares
1 square foot	0.09 square metres
1 pint	568 millilitres

Quiz 61

1) 33.8°F
2) –17.2°C
3) 68°F
4) 15°C
5) 23°F
6) –25°C
7) 279°K
8) –135°C
9) 298°K
10) 41°F

Quiz 62

1) Tommy's
2) The red sports car
3) The blue corner
4) Farmer Silage
5) The puce room
6) New York

Quiz 63

1) Mean: 6.6; Median: 7.5; Mode: 8
2) 8°
3) Negative correlation
4) One billion (1,000,000,000)
5) 10kg of potatoes
6) Tony
7) Geraldine
8) 64.4°F

Quiz 64

1) Equation
2) Formula
3) Equation
4) Equation
5) Formula
6) Equation

Quiz 65

1) 3b and 21
2) 14, –6w, 5w, –8
3) 6 and 5 are coefficients; w is a variable
4) t and 2t; –4 and 2
5) 2p and p^2; 2p and 20; p^2 and 20; 3p and 20 are all pairs of unlike terms

Quiz 66

1) a = 18

   a + 7 – 7 = 25 – 7 = 18

2) c = 16

   28 – c + c = c + 12 ⇨ 28 – 12 = c + 12 –12 ⇨ 16 = c

3) d = 9

   8 + d + 2 = 19

   d = 19 – (8 + 2) = 19 –10 = 9

4) e = 5

   48 – 11 – e + e = 32 + e ⇨ 48 – 11 – 32 = 32 + e – 32 ⇨ 5 = e

   e = 48 – 32 – 11 = 5

5) f = 16

   19 + f – 7 + 7 = 28 + 7 ⇨ 19 + f = 35 ⇨19 + f – 19 = 35 – 19 ⇨ f = 16

6) g = 54

   24 + 36 – 5 – g + g = 1 + g ⇨ 24 + 36 – 5 – 1 = 1 + g – 1 ⇨ 24 + 36 – 5 – 1 + g ⇨ 54 = g

7) h = 57

   112 – h + 43 + h = 7 x 14 + h ⇨ 112 + 43 = 98 + h ⇨ 155 = 98 + h ⇨ 155 – 98 = h ⇨ 57 = h

8) i = 19

   45 ÷ 9 + i = 72 ÷ 3 ⇨ 5 + i = 72 ÷ 3 ⇨ 5 + i = 24 ⇨ i = 24 – 5 ⇨ i = 19

Quiz 67

1) $h = 8$

   $3h \div 3 = 24 \div 3 \Rightarrow h = 24 \div 3 = 8$

2) $i = 9$

   $7i - 14 + 14 = 49 + 14 \Rightarrow 7i = 63 \Rightarrow 7i \div 7 = 63 \div 7 \Rightarrow i = 63 \div 7 = 9$

3) $j = 99$

   $j/3 \times 3 = 33 \times 3 \Rightarrow j = 33 \times 3 = 99$

4) $k = 144$

   $k/12 + 38 - 38 = 50 - 38 \Rightarrow k/12 = 50 - 38 \Rightarrow k/12 = 12 \Rightarrow k/12 \times 12 = 12 \times 12 \Rightarrow k = 12 \times 12 = 144$

5) $e = -7$

   $3e + 11 - 11 = -20/2 - 11 \Rightarrow 3e = -10 - 11 = -21 \Rightarrow e = -21 \div 3 = -7$

6) $f = -6$

   $3f \div 9 \times 9 = -2 \times 9 = -18 \Rightarrow f = -18 \div 3 = -6$

Quiz 68

1) $3a + 1 = 7$
2) $2b^2 + 7b = 22$
3) $c^4 3^4 = 7{,}776$
4) $4d^5 = 972$
5) $e^4 = 81$
6) $2f^2 = 18$

Quiz 69

1) a = 3

   (2 x a) + (2 x 4) = 14 ⇨ 2a + 8 = 14 ⇨ 2a = 14 – 8 = 6 ⇨ a = 6/2 = 3

2) b = 13

   (6 x b) + (6 x –6) = 42 ⇨ 6b –36 = 42 ⇨ 6b = 42 + 36 = 78 ⇨ b = 78/6 = 13

3) c = 7

   (4 x c) + (4 x 3) = 32 ⇨ (4c + 12) – 8 = 32 ⇨ 4c + 12 = 42 + 8 = 40 ⇨ 4c = 40 – 12 = 28 ⇨ c = 28/4 = 7

4) d = 9

   (–3 x d) + (–3 x 4) = –39 ⇨ –3d – 12 = –39 ⇨ –3d = –39 + 12 = –27 ⇨ d = –27/–3 = 9

5) e = –7

   (3 x e) + (3 x 2) = –15 ⇨ 3e + 6 = –15 ⇨ 3e = –15 – 6 = –21 ⇨ e = –21/3 = –7

6) f = –4

   (–6 x f) + (–6 x –3) = 42 ⇨ –6f + 18 = 42 ⇨ –6f = 42 – 18 = 24 ⇨ f = 24/–6 = –4

Quiz 70

1) $a = 4$

   $6a + 2 = 4a + 10 \Rightarrow 2a + 2 = 10 \Rightarrow 2a = 8 \Rightarrow a = 4$

2) $b = 7$

   $12b - 5 = 10b + 9 \Rightarrow 2b - 5 = 9 \Rightarrow 2b = 14 \Rightarrow b = 7$

3) $c = 9$

   $3c + 22 = 6c - 5 \Rightarrow 22 = 3c - 5 \Rightarrow 27 = 3c \Rightarrow 9 = c$

4) $d = 5$

   $2(d + 2) = 3d - 1 \Rightarrow 2d + 4 = 3d - 1 \Rightarrow 4 = d - 1 \Rightarrow 5 = d$

5) $e = 23$

   $e + 49 = 3(e + 1) \Rightarrow e + 49 = 3e + 3 \Rightarrow 49 = 2e + 3 \Rightarrow 46 = 2e \Rightarrow 23 = e$

6) $f = -2$

   $3f + 2 = 4f + 4 \Rightarrow 2 = f + 4 \Rightarrow -2 = f$

7) $g = -7$

   $18 - g = -3g + 4 \Rightarrow 18 = -2g + 4 \Rightarrow 14 = -2g \Rightarrow -7 = g$

8) $h = 14$

   $4h - 5 = 3(h + 3) \Rightarrow 4h - 5 = 3h + 9 \Rightarrow h - 5 = 9 \Rightarrow h = 14$

Quiz 71

1) 35

   $a + 10 = 45$ ⇨ $a = 45 - 10 = 35$

2) 42

   $3b = 126$ ⇨ $b = 126/3 = 42$

3) 78

   $c/3 + 14 = 40$ ⇨ $c/3 = 40 - 14 = 26$ ⇨ $c = 26 \times 3 = 78$

4) 79

   $(d + 5)/3 = 28$ ⇨ $d + 5 = 28 \times 3 = 84$ ⇨ $d = 84 - 5 = 79$

5) 19

   $3e + 17 = 4e - 2$ ⇨ $17 = e - 2$ ⇨ $e = 17 + 2 = 19$

Quiz 72

1) $8(8 + 3) = 88$
2) $3(3a + 5) = 51$
3) $12(2b - 3) = 12$
4) $c(c + 5) = 66$
5) Cannot be factorized
6) $3f(f + 2) = 72$

Quiz 73

1) Quadratic
2) Not quadratic
3) Quadratic
4) Not quadratic
5) Quadratic

Quiz 74

1) $3x^2 + 6x - 7 = 0$
2) $x^2 + 3x - 2 = 0$
3) $x^2 - 4x + 2 = 0$
4) $3x^2 - 9x - 8 = 0$
5) $x^2 + 4x + 6 = 0$

Quiz 75

1) 3 and 7
2) 6 and 9
3) –3 and 2
4) –4 and 8
5) 5 and 44
6) 28 and 19

Quiz 76

1) a = 5
   (a – 5) (a – 5)

2) b = –7 or b = –13
   (b + 7) (b + 13)

3) c = –2 or c = 4
   (c + 2) (c – 4)

4) d = 6 or d = –9
   (d – 6) (d + 9)

5) e = –4 or e = –17
   (e + 4) (e + 17)

Quiz 77

1) a = 5; b = 7
2) c = 9; d = 6
3) Jack is 48; Beryl is 36
4) Pete earns £6.50 per hour; Sylvia earns £7.00 per hour
5) A sandwich costs £1.40; a chocolate bar costs £0.75

Quiz 78

1) a = 24
2) b = 18
3) c = 42
4) d = 17
5) e = 29
6) f = –6
7) g = –5 or g = –3
8) A cup of tea costs 85 pence and a hot chocolate costs £1.35

Quiz 79

1) Right angle	iii)	An angle of 90°	B)
2) Acute angle	i)	An angle less than 90°	D)
3) Obtuse angle	iv)	An angle between 90° and 180°	C)
4) Reflex angle	ii)	An angle greater than 180°	A)

Quiz 80

Number of Sides	Name of Polygon
3	Triangle
4	Quadrilateral
5	Pentagon
6	Hexagon
7	Heptagon
8	Octagon
9	Nonagon
10	Decagon
11	Hendecagon
12	Dodecagon
20	Icosagon
50	Pentacontagon
100	Hectagon

Quiz 81

1) c) Equilateral triangle
2) f) Rectangle
3) g) Kite
4) a) Isosceles triangle
5) h) Square
6) b) Rhombus
7) e) Parallelogram
8) i) Trapezium
9) d) Scalene triangle

Quiz 82

1) 90°
2) 108°
3) 120°
4) 128.571°
5) 135°
6) 140°
7) 144°
8) 147.273°
9) 150°
10) 162°

Quiz 83

1) a = 115°
2) b = 60°
3) c = 110°
4) d = 110°
5) e = 130°

Quiz 84

1) 40cm
2) 117cm
3) 120cm
4) 17cm
5) Dodecagon
6) 144cm

Quiz 85

1) $49cm^2$
2) $15cm^2$
3) $80cm^2$
4) $54cm^2$
5) $240cm^2$
6) $209cm^2$
7) $108cm^2$

Quiz 86

1) h = 13 metres
- $h^2 = 12^2 + 5^2 = 144 + 25 = 169$
- h = square root of 169 = 13

2) b = 3 metres
- $b^2 = h^2 - a^2 = 5^2 - 4^2 = 25 - 16 = 9$
- b = square root of 9 = 3

3) a = 8 metres
- $a^2 = h^2 - b^2 = 10^2 - 6^2 = 100 - 36 = 64$
- a = square root of 64 = 8

4) Area = 117 metres^2
- $h^2 = a^2 + b^2 = 6^2 + 9^2 = 36 + 81 = 117$

Quiz 87

1) Circumference	c) The perimeter of a circle
2) Radius	a) The distance from the centre of a circle to its edge
3) Diameter	d) The distance across the whole of a circle via its centre point
4) Pi	b) A value (beginning 3.14159) key to calculating circumference

Quiz 88

1) 11.75cm
2) 118.54cm
3) 87.96cm
4) 3,848.45cm^2
5) 7,013.8cm^2
6) 12cm

Quiz 89

1) c) Sphere
2) g) Cube
3) f) Square pyramid
4) e) Cylinder
5) b) Tetrahedron
6) h) Cuboid
7) j) Torus
8) i) Cone
9) d) Pentagonal prism
10) a) Octahedron

Quiz 90

1) $729cm^3$
2) $904.78cm^3$
3) $75cm^3$
4) $240cm^3$
5) $2,035.75cm^3$
6) $942.48cm^3$

Quiz 91

1) Tetrahedron
2) Cuboid
3) Square pyramid
4) Triangular prism
5) Cylinder
6) Cone

Quiz 92

Quiz 93

1) 6.43cm

   Using sin θ = $^{opposite}/_{hypotenuse}$, sin 40 = $^{a}/_{10}$ ⇨ 0.6428 = $^{a}/_{10}$. Therefore, a = 0.6428 x 10 = 6.428 = 6.43cm

2) 6.93cm

   Using cos θ = $^{adjacent}/_{hypotenuse}$, cos 30 = $^{b}/_{8}$ ⇨ 0.8660 = $^{b}/_{8}$. Therefore, b = 0.8660 x 8 = 6.928 = 6.93cm

3) 8.4cm

   Using tan θ = $^{opposite}/_{adjacent}$, tan 55 = $^{12}/_{c}$ ⇨ 1.4281 = $^{12}/_{c}$. Therefore, c = $^{12}/1.4281$ = 8.4027 ⇨ 8.4cm

4) 42°

   Using sin θ = $^{opposite}/_{hypotenuse}$, sin d = $^{6}/_{9}$ = 0.6666 ⇨ d = inv. sin of 0.6666 = 41.8052 ⇨ 42°

5) 54°

   Using cos θ = $^{adjacent}/_{hypotenuse}$, cos e = $^{7}/_{12}$ = 0.5833 ⇨ e = inv. cos of 0.5833 = 54.3147 ⇨ 54°

Quiz 94

1 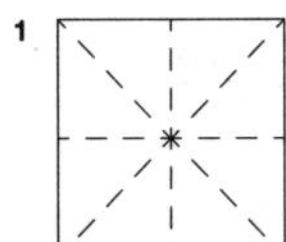

2 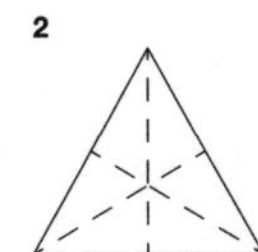

3 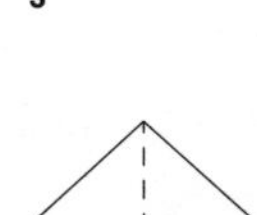

4 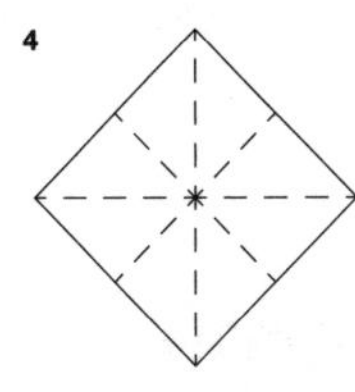

5 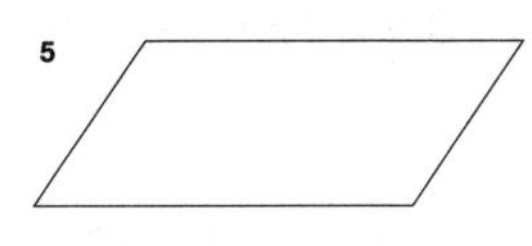

6 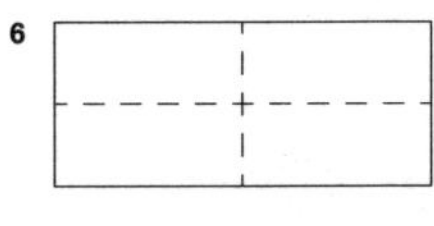

7 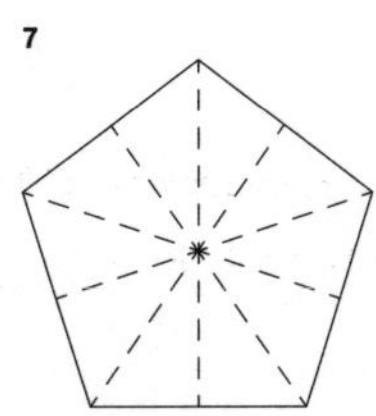

8 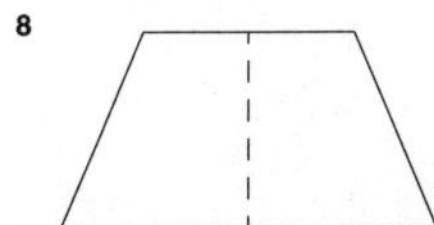

1) 4 axes of symmetry
2) 3 axes of symmetry
3) 4 axes of symmetry
4) 4 axes of symmetry
5) 0 axes of symmetry
6) 2 axes of symmetry
7) 5 axes of symmetry
8) 1 axis of symmetry

Quiz 95

1) Reflection
2) Translation
3) Rotation (clockwise)
4) Rotation (clockwise)
5) Translation
6) Reflection
7) Rotation (anti-clockwise)
8) Rotation (clockwise)

Quiz 96

1) A pentagon
2) 56cm
3) 192cm^2
4) 12cm
5) 40.84cm
6) 4,800cm^3
7) 1,244cm^3
8) 37°

Quiz 97

1) 40, 43, 44, 45, 49, 60, 60, 61, 62, 66
2) a = 17; b = 10
3) c = 19; d = 23
4) 168, 180, 210, 448, 560, 560, 672, 700, 720, 840
5) Route C

Quiz 98

1) £580
2) 1 hour 28 minutes
3) 7 times
4) 132cm
5) 15 discs
6) 705 runs
7) 25 people
8) 6 boxes

Quiz 99

(1) 1	3	2	■	(3) 1	(4) 2
(5) 4	3	■	(6) 1	0	8
(7) 4	■	(8) 1	4	1	■
■	(9) 2	7	■	■	(10) 5
(11) 1	2	■	(12) 1	5	6
(13) 6	6	3	■	(14) 2	4

Quiz 100

1) 132
2) 136 sandwiches
3) 180
4) 10100000 (5 x 32 = 160)
5) 1,459 books
6) 27.5
7) 42°
8) $\frac{9}{20}$
9) £2,387
10) 13
11) 15 cm
12) XCII (179 – 87 = 92)
13) 66 and 130 (multiply the previous figure by 2 and then deduct 2)
14) $11\frac{1}{2}$
15) a = 7
16) 0.67
17) 22cm
18) £11,505
19) $\frac{15}{231}$
20) Mean: 154 days; Median: 59 days; Mode: 14 days
21) 43.7
22) £264.60
23) m = 19; n = 9
24) 459cm
25) 83

NOTES

NOTES

NOTES

NOTES